James Brindley

and the early engineers

Lives of the Engineers
Vermuyden - Myddelton - Perry - Brindley

Samuel Smiles

"Bid Harbours open, Public Ways extend;
Bid Temples, worthier of God, ascend;
Bid the broad Arch the dang'rous flood contain,
The Mole projected break the roaring main,
Back to his bounds their subject sea command,
And roll obedient rivers through the land.
These honours, Peace to happy Britain brings;
These are imperial works, and worthy kings."

POPE.

TEE Publi
Warwickshire

Cornelius Jansen   William Holl

Sir Hugh Myddelton, Knight & Baronet

First published 1874
Reprinted 1999
© TEE Publishing
ISBN 1 85761 110 1

INTRODUCTION,

LORD BACON hath truly said that there are three things which make a nation great and prosperous,—a fertile soil, busy workshops, and easy conveyance for men and commodities from one place to another.

The soil of England is sufficiently fertile; but it has been made so principally by the industry of its people. A large portion of it, and that the most fertile, has been reclaimed and embanked from the sea. The two thousand square miles of land in Lincolnshire and Cambridgeshire, forming the Great Level of the Fens, would, but for human skill and industry, be lying under water, the haunt of wild birds.

Who does not know of the business of our Workshops? —of the employment they give to large numbers of the population; of the tools they fabricate; of the articles of commerce they produce; of the demands they supply for consumers in all parts of the world.

Then, with respect to Lord Bacon's "easy conveyance for men and commodities from one place to another:" have we not our Roads, our Canals, and our Railways; and has not England also been the originator of the Steam engine, the Steamboat, and the Screw ship?

a 2

Although the inventions of British engineers now exercise so great an influence all over the world, it may possibly excite some surprise to find how very recently England has taken up her present position in regard to Engineering. Indeed, skilled Industry may be said to form the youngest outgrowth of our national life.

Most of the continental nations had a long start of us in art, in science, in mechanics, in navigation, and in engineering. In former times we were more given to fighting than working. The wars between England and France—between England and Scotland—between England and Ireland—deterred our people from engaging in the pursuits of industry and commerce. The only article we exported was wool. We seem to have been unable to make our own clothing; for a large part of the exported wool was returned to England as cloth, for sale in the English markets. We grew wool for Flanders, as America grows cotton for England now. Even the small quantity of cloth manufactured at home was sent to the Low Countries to be dyed.

Many branches of industry have been established in England by foreigners. The Flemings introduced cloth-making, stuff-making, and linen-weaving. The brothers Elers, Dutchmen, began the pottery manufacture. Spillman, a German, established the first paper-mill at Dartford; and Bromen, another Dutchman, brought the first coach into England. Towards the end of the seventeenth century, a large number of Huguenot manufacturers established various branches of industry in England and Ireland which, until then, had for the most part been conducted abroad.

We owe the first working of our mines to the Germans; who brought with them nearly all the mining terms still in use among us. Elizabeth invited skilled miners from Germany to settle in different parts of England, for the purpose of teaching our people the best methods of working the ores. Two of these, Hochstetter and Thurland, of Augsburg, established copper-works at Keswick, which were worked to such advantage that the Queen said she had left more brass than she had found iron ordnance in England.

Not many centuries since, Italy, Spain, France, and Holland, looked down contemptuously on the poor but proud islanders, contending with nature for a sustenance amidst their fogs and their mists. Though surrounded by the sea, we had scarcely any Royal navy until within the last three hundred years. An old writer, in his eulogium of Henry VII., said: "He built a chapel; he built a ship." But we had no Commercial navy. The carrying trade of the world was then performed by the Dutch, the Genoese, the Venetians, and the Portuguese.

When we wanted any skilled work done, we almost invariably sent abroad for foreigners to do it. Our first ships were built by Danes, Venetians, and Genoese. When the royal ship, the 'Mary Rose,' sank at Spithead in 1545, a company of Venetians was hired to raise her. On that occasion Peeter de Andreas was employed, assisted by his ship carpenters and sailors, with " sixty English maryners to attend upon them."

Our first lessons in mechanical and civil engineering were principally obtained from the Flemings and Dutchmen, who supplied us with our first wind-mills, water mills, and fulling-mills. When an engine was required to pump water from the Thames, Peter Morice, the Dutchman, was employed to erect it. The first iron cannons were cast at Buxted, in Sussex, by Peter Baude, a Frenchman; and the first mortar pieces by Peter Van Collet, a Flemish gunsmith.

Holland even sent us the necessary engineers and labourers to execute our first great works of drainage. The Great Level of the Fens was drained by Vermuyden, a native of Zealand. Another Dutchman, Freestone, was employed to reclaim the marsh, near Wells, in Norfolk. Canney Island, near the mouth of the Thames, was embanked by Joas Croppenburgh and his company of Dutch workmen.

When a new haven was required at Yarmouth, Joas Johnson, the Dutch engineer, was employed to plan and construct the works. When Dover habour was threatened with ruin in the reign of Henry VIII., Ferdinand Poins, a Fleming, was invited to come over to plan a new quay for the purpose of preserving the haven. Matthew Hake, of Gravelines, in Flanders, was also sent for about the same time, for the purpose of repairing a serious breach which had been made in the banks of the river Witham, at Boston; and he brought with him not only the mechanics but the manufactured iron required for the work.

Indeed very little English iron was then used. Almost the only iron in England was imported by the Foreign Merchants of the Steelyard Company. The Spaniards prided themselves upon the superiority of their armour, and regarded the scarcity of iron in England as an important element of success in their invasion of this country by the great Armada. Down to the end of last century, England was principally indebted for its iron to Sweden, Germany, Spain, and Russia.

Yet, a very long time ago, there must have been great architects as well as great builders in England. This is sufficiently proved by the magnificent cathedrals, abbeys, and ancient bridges which still exist. It is not improbable that these old architects and builders were Frenchmen, the western part of France being for many hundred years subject to the English crown. However that may be, the English architects and builders must have greatly fallen off by the time of Elizabeth. Thus, when Sir Thomas Gresham built the Royal Exchange in 1566, he brought from Flanders the requisite masons and carpenters to execute it under the direction of Henryke, their master builder. The foreigners also brought with them all necessary materials—the wainscot, the glass, the slates, the iron, and even much of the stone for the building. In short, as Holinshed relates, Gresham " bargained for the whole mould and substance of his workmanship in Flanders." The only Englishmen employed upon the work, were the labourers who carried the lime and stone required for the building.

Bridge building also must have greatly fallen off. Little more than a hundred years ago, there were few architects or masons able to build a bridge of any extent : and when a second bridge had to be erected over the Thames at London, the French engineer Labelye, a native of Switzerland, was sent for to build it; though the original Westminster Bridge is already destroyed, and replaced by a new structure.

In short, we depended for our engineering, even more than we did for our pictures or our music, upon foreigners. At a time when Holland had completed its magnificent system of water communication, and when France, Germany, and even Russia, had opened up important lines of inland communication, England had not cut a single canal, while our roads were about the worst in Europe.

The first English canal was made by a workman taken from the lowest rank of society. He could not write his own name. But he was a persevering, ingenious man, possessed of a shrewd mother wit. Brindley would have become great in his own trade—that of a millwright; but having succeeded in accomplishing several important works which he had undertaken, he attracted the attention of the Duke of Bridgewater, who employed him to make the first English canal between Worsley and Manchester.

Water was the great difficulty of the early Engineer. First he had to dam it out or bale it out, like Vermuyden. Or he erected a lighthouse upon a rock in the ocean, to withstand its power, like Smeaton. Or he founded piers, bridges, and docks, in the midst of it, like Rennie. The Engineer had also to make water serve his purposes He led it into a mill-stream, and made it grind his corn and work his machinery. Or he collected it into reservoirs to supply a large city with water, like Myddelton; or, like Brindley, he made it fill long canals, extending between one town and another, and thus made a great water road to convey coal, and minerals, and merchandise. Thus, in the hands of the Engineer, water, instead of being a tyrant, became a servant; instead of being a destroyer, it became a useful labourer and a general civiliser.

Richard Cobden has said, that "the opening up of the internal communications of a country, is undoubtedly the first and most important element of its growth in commerce and civilisation. Hence our canals were regarded by Baron Dupin as the primary material agents of the wealth of Great Britain." Roads, however, are more important than canals; and at the time when the first canals were made, the roads of England still remained in a deplorable condition.

Down to the middle of the last century, it was almost impossible to travel with speed or comfort in any direction. Everything was carried on horse-back, and sometimes on bullock-back. Corn, coal, wool, iron, and such like articles, were carried through the country on pack-horses. The roads were not roads so much as tracks ; and when the old tracks became dangerous through depth of mud, new tracks were struck out across the adjoining fields. Guides were employed to keep travellers out of the mud ; for it was often thick and slab. During the civil war, some eight hundred horse were taken prisoners while sticking in the mud in Buckinghamshire.

Where all merchandise was carried by pack-horses over bad roads, domestic commerce was simply impossible. Hence the commerce of London was mostly Foreign. We bought cloth from Belgium, silk from France, cutlery from Italy, hats from Flanders, Delftware from Holland, and iron from Spain, Sweden, and Russia. It cost so much less to bring goods from Hamburg, Amsterdam, or Havre by sea, than from Norwich, Birmingham, or Manchester by land.

The English highroads were also a good deal beset by highwaymen. The traveller often passed along bridlo ways through fields, where frequent gibbets warned him of his perils.

When the roads were attempted to be made suitable for carts, waggons, and carriages, they were very little improved. The ston or metal was put on, and the roads were simply left to manage themselves. They soon became mere slumpers of stones and mud. If they were mended at all, it was only by throwing big stones into the biggest holes. Arthur Young when travelling in the northern counties of England, in 1768, measured the depth of the tracks he travelled through. Between "proud Preston" and Wigan, he says, "I actually measured ruts of four feet deep, floating with mud only from a wet summer; and between the towns I actually passed three carts broken down in those eighteen miles of execrable memory."

In another place, he says, "You will form a clear idea of these roads, if you suppose them to represent the roofs of houses, and the roads to run across them." In winter, he says, it would have cost no more money to make the roads *navigable*, than to make them hard. Even near London, the roads were often impassable gulfs of mud. It then took a longer time for a carriage to reach twenty miles out of London, than it now does to reach York or Manchester.

An Act was passed in 1663, authorizing the first turnpike roads to be made, with the usual toll-gates; but the people would not bear the tolls, and the gates were pulled down wherever erected. As late as 1753, a general raid was made upon the toll-gates in Yorkshire. About a dozen were destroyed weekly. The soldiers were at length sent to protect the turnpikes; and on their apprehending a score of the rioters to convey them to York, the people attempted a rescue, when the soldiers fired, and killed and wounded a considerable number.

Nevertheless, considerable progress was made in the improvement of the roads, especially after the last Highland rebellion. Between 1760 and 1780, about six hundred Acts were passed, authorising the construction of new roads and bridges; and by the end of the century, the main roads were in a very fair condition.

The early roads of Scotland and Wales were even worse, if possible, than those of England. Sometimes they lay along the bed of a river. At other times they crossed a barren mountain. In the lower grounds, bogs were always in the way. The Life of Telford will show what that able engineer did towards opening up Scotland and Wales, and making the roads passable, for pleasure or for business, at all times of the year.

The construction of the Highland roads had also an important moral influence. The roads stimulated industry amongst a people who had hitherto been unused to it. In constructing them, the people learnt to work, to use tools, and to apply themselves to continuous labour. Telford himself regarded the Highland roads in the light of a Working Academy, which annually turned out about eight hundred improved workmen.

Down to about the middle of last century, the Highlanders were among the most turbulent of subjects. When not lifting cattle from the Lowlanders, they were engaged in arms against the Government. They levied black-mail all round the Highland frontier as late as 1745. Mr. Grant, of Corrymorry, a Highland chief, used to insist that " a rising in the Highlands was absolutely necessary to give employment to the numerous bands of lawless and idle young men who infested every property."*

During the rebellion of 1745, there was a company of men in one of Prince Charles's regiments, called " The Balmoral Farquharsons." These men fought to the best of their power against her Majesty's forefathers. They were raised from the estate on which the Queen now resides; and perhaps she has not more peaceable nor affectionate subjects than the Balmoral men, nor than the Highlanders generally, in all her dominions.

After the turnpike roads had been completed—after the hills had been cut down and the valleys raised, so as to level the roads in all directions—after the mails were able to travel at ten miles an hour from London almost to John O'Groat's—and when all complaints as to the quickness of travelling seemed to be removed,—lo! the Locomotive was invented, and all that Telford had done was, to a certain extent, undone by the Stephensons.

One of the last works of Telford was the improvement of the high road between London and Birmingham, so as to enable the coach-road to compete with the rail-road. He was levelling the top of the hill at Weedon, near Daventry, while the railway excavators were working underneath. Near that point, three remarkable things are to be seen,— the remains of Watling Street, the original Roman road,— Telford's last turnpike road,—and the first great rail-road between London and the north.

It will thus be seen, that notwithstanding our delays in opening up our internal communications, we have at

---

* Anderson's ' Highlands and Islands of Scotland,' p. 434.

last effectually made up our long arrears. We have no longer to send for foreign engineers to execute our necessary works in Embankments and Bridge building. Our native engineers can make docks and harbours without extraneous assistance. We can make our own iron, and our own tools. In fact, instead of borrowing engineers from abroad, we now send them to all parts of the world. British built ships ply on every sea. We export machinery to all countries, and supply Holland itself with pumping engines.

During the last hundred years, our native engineers have completed a magnificent system of canals, turnpike roads, bridges, and railways, by means of which the internal communications of the country have been completely opened up. They have built lighthouses—finger-posts of the sea—round our coasts, by which ships freighted with the produce of all lands are safely lighted to their destined havens. They have hewn out and built docks and harbours for the accommodation of a gigantic commerce; whilst their inventive genius has rendered iron, fire, and water, the most untiring workers in all branches of industry, and the most effective agents in locomotion by land and sea.

The remarkable thing is, that nearly all this has been accomplished during the last hundred years, and much of it within the life of the present generation.

Smeaton did not erect and light up the first stone lighthouse in Britain, until the year 1759.

Brindley made the first canal for the Duke of Bridgewater in 1761.

Although Watt had been engaged for about ten years

in inventing his condensing steam-engine, it was not until 1776 that he erected and sold his first engine. It was patented in 1769, the same year in which Arkwright patented his machine for spinning by rollers.

It was not until the end of last century that England was able to dispense with the supply of the greater part of its iron from abroad; for Henry Cort did not invent his puddling process, which so enormously increased the production of iron, until the years 1783 and 1784.

The first mail coach, on Palmer's system, did not begin to run between London and Bristol until the year 1784.

There were no public Docks in London until the beginning of the present century. The first were constructed by Jessop and Rennie.

The first public exhibition of gas-lighting was made in 1802, when Murdock lit up the front of the Soho manufactory with gas, in celebration of the Peace of Amiens.

Trevithick used his first high-pressure locomotive in 1802, when he ran it as a road locomotive along the streets of his native town.

1802 was an eventful year, for it was then that Symington employed his first successful steamboat on the Forth and Clyde canal.

Fulton's steamboat, with Watt's engine as the motive power, made its first voyage on the Hudson in 1807; and Henry Bell's 'Comet' made its first voyage on the Clyde in 1812; but a considerable time elapsed before steam was generally introduced into either the Commercial or the Royal navy.

In 1819, the 'Savannah' steamboat, of 350 tons, made the

voyage from New York to Liverpool in twenty-six days; and in 1838, the 'Great Western' arrived at New York from Bristol in eighteen days. Since then the Atlantic Ocean has been covered with screw steam-ships; and the direct line between Liverpool and New York has become so crowded, that steam-ships must keep a course so definite that the expression of " steam lanes " has become a matter of ordinary maritime parlance.

George Stephenson invented his Killingworth locomotive for drawing coals in 1814; and his passenger locomotive for the Liverpool and Manchester Railway in 1829.

The first telegraph line was erected on the Great Western Railway, between Paddington and West Drayton, in 1841; and the submarine telegraph was first laid between Dover and Calais in 1851. Since then the telegraph has really " put a girdle round the earth."

The following comprehensive thought occurs in one of Sir Isaac Newton's Reflections :—

" It is clearly apparent that the inhabitants of the world are of a short date, seeing that all arts, as letters, ships, printing, the needle, and such like, have been discovered within the memory of history."

The discoveries of Newton himself were among the greatest of the age he lived in; but since his time various others have been made which strikingly illustrate the truth of his observation.

Only thirty years after his death, Harrison invented the Chronometer, by means of which the navigator can determine his longitude at sea,—an invention almost equal to that of the Mariner's Compass itself.

The science of Chemistry has been discovered,—dating

from the days of Black, Priestley, and Lavoisier; and Natural Philosophy, if not discovered, has at least been greatly developed since Newton's time; while both have been the subject of many inventions of the greatest importance.

Thus the inventions of the steam-engine, the locomotive, and the electric telegraph, have already wonderfully affected the civilized world; and are probably destined to accomplish still more extraordinary changes in the generations that are to come.

One of the most remarkable things about Engineering in England is, that its principal achievements have been accomplished, not by natural philosophers nor by mathematicians, but by men of humble station, for the most part self-educated.

The educated classes of the last century regarded with contempt mechanical men and mechanical subjects. Dean Swift spoke of "that fellow Newton over the way—a glass grinder, and a maker of spectacles." Smeaton was taken to task by his friends of the Royal Society, for having undertaken the "navvy" work of making a road across the valley of the Trent, between Markham and Newark.

At a time when the Court, the Camp, and the Church, formed the principal occupations of the higher classes, engineering was thought unscientific and ungenteel. Hence, when any public work of importance had to be done—some great level of the Fens to be drained, or some harbour to be constructed—a foreign engineer was usually employed to do it.

Nor did any of the great mechanics, who have since

invented tools, engines, and machines, at all belong to the educated classes. They received no college education. Some of them could scarcely write their own names. But where learning failed, natural genius triumphed. These men gathered their practical knowledge in the workshop, or acquired it in manual labour. They rose to celebrity, mostly by their habits of observation, their powers of discrimination, their constant self-improvement, and their patient industry.

The greatest stimulus to English engineering has been Trade—the increase of our commerce at home, and the extending of it abroad. England was nothing, compared with continental nations, until she had become commercial. She fought wonderfully, and boasted of many victories; but she was gradually becoming less powerful as a State, until about the middle of last century, when a number of ingenious and inventive men, without apparent relation to each other, arose in various parts of the kingdom, and succeeded in giving an immense impulse to all the branches of the national industry; the result of which has been a harvest of wealth and prosperity, perhaps without a parallel in the history of the world.

Before that time, our commerce had grown very slowly. Spain and Portugal had immense foreign possessions and commerce, when England had none. A small company of English merchants began to trade with India about the beginning of the seventeenth century. The northern colonies of America were founded about the same time; but the government of England afterwards threw them off. Admiral Penn took Jamaica from the Spaniards in 1655; but the other principal islands and possessions in the West Indies were only ceded to England during the wars of the French Revolution. We obtained possession of the Cape of Good Hope in 1815. New South Wales colony began to be planted in 1788, South Australia in 1836, and Victoria in 1837. The New Zealand Company was only established in 1839.

With the increase of her commerce, about the end of last century, England became more "quick" than she had been before. Intercourse between town and town became more frequent, for the purposes of trade. Then the roads became improved; canals were made; bridges were built; docks and harbours were constructed; and, above all, the steam-engine was invented. Within a few years, Watt's invention created a large number of new industries, and gave employment to immense numbers of people. The engine was perfected while England was engaged in the Revolutionary war with France; and it not only kept our armies in the field, paid the interest upon our debt, but left the nation more powerful than ever it had been before.

At a meeting held in London in 1824, for the purpose of erecting a monument to James Watt,—the greatest of engineers,—Mr. Huskisson used these remarkable words:

"His invention has contributed not only to advance personal comfort and public wealth, by affording to industrious millions the faculty of providing for their industrial wants, by means which directly conduce to the general power and greatness of the state, but also to the general diffusion of a spirit of improvement, a thirst for instruction, and an emulation to apply it to the purposes of practical utility, even in the humblest classes of the community. . . . Looking back, however, to the demands which were made upon the resources of the country during the late war, perhaps it is not too much to say, at least it is my opinion, that these resources might have failed us before that war was brought to a safe and glorious conclusion, but for the creations of Mr. Watt, and of others moving in the same career, by whose discoveries those resources were so greatly multiplied and increased. It is perhaps not too much to say, that but for the vast accession thus imperceptibly made to the general wealth of the empire, we might have been driven to sue for peace, before, in the march and progress of events, Nelson had put forth the last energies of his naval genius at Trafalgar, or, at any rate, before Wellington had put the final seal to the security of Europe at Waterloo."

Watt, however, did far more than assist in the conquests of Trafalgar and Waterloo. He discovered an immense amount of Latent Labour. He yoked his steam-engine to machinery of almost infinite variety, and employed it to create materials of vast importance for the comfort of mankind. The steam-plough and the steam-press are alike its issue. It impels ships, drives locomotives, hammers iron, excavates docks, and, in a word, is the greatest and most indefatigable worker in the world. Even the steam-engine itself was never complete until it could make itself—that is, manufacture the self-acting tools which could never go wrong in the process of constructing it.

One of the circumstances that has particularly attracted the attention of foreigners has been, that almost the whole of our great engineering works have been the spontaneous outgrowths of private and individual industry and genius. Even Watt was a mere tradesman, trying for ten years to work out his great invention; and, but for the noble and generous help of Matthew Boulton, he might never have perfected it.

Government has done next to nothing to promote engineering works. These have been the result of the liberality, public spirit, and commercial enterprise of merchants, traders, and manufacturers. Baron Dupin, in his review of the engineering works of England, observes that "fewer years sufficed for a few individuals to execute and construct, at their private expense, the docks of England, which receive the trading fleets of the two hemispheres, than it required for the triumphant government of France to erect some of the quays of the Seine. These are truly the prodigies of the seas."*

Canova is said to have wondered, while in England, that the trumpery Chinese bridge, then in St. James's Park, should have been the production of the Government, while Waterloo Bridge was the enterprise of a private company. Baron Dupin, speaking of Rennie's masterpiece, uses these words :—

"If, from the incalculable effect of the Revolutions to which empires are subject, the people of the earth should one day inquire, ' Which was formerly the New Phœnicia and the Western Tyre, which covered the ocean with her vessels?'—The greater part of her edifices, consumed by a destructive climate, will no longer stand to answer with the dumb language of monuments; but Waterloo

---

* 'Mémoires sur la Marine.' 1819.

Bridge will ever exist to repeat to the most remote generations, 'Here stood a rich, industrious, and powerful city!' The traveller at this sight will imagine that some great prince sought to signalise the end of his reign by many years of labour, and to immortalise the glory of his actions by this imposing structure. But if tradition tell him that six years sufficed to begin and complete this work—if he learn that a mere company of merchants built this mass, worthy of Sesostris and the Cæsars—he will the more admire the nation where similar enterprises could be the fruit of the efforts of merchants and private individuals. And if he should then reflect on the causes of the prosperity of empires, he will understand that such a nation must have possessed wise laws, powerful institutions, and a well-protected liberty; for these are stamped in the grandeur and utility of the works completed by her citizens."*

The incidents connected with the Lives of the Engineers themselves are of much interest. In the first place, the early engineers thoroughly represented the national character. Baron Dupin took this view. He said we could not understand the astounding successes of English commerce and the sustained skill and enterprise of English engineers, without taking into consideration that perseverance in all undertakings and that constancy of character on which their successes mainly depended.

There is a promptitude about the Engineer that we do not find in the Architect. The late Prince Consort is reported to have said: " If we want any work done of an

* Baron Dupin, 'On the Commercial Power of Great Britain, Vol. i. p. 360. 1825.

unusual character, and send for an architect, he hesitates, debates, trifles : we send for an engineer, and he *does it*." Hence the Prince usually relied on Engineers for carrying out the important works with which he was connected.

The greater number of the early engineers were the offspring of necessity. Some great work required to be done, and forthwith a skilled worker (for there were no so-called engineers in those days) was called upon to do it. The work which he had undertaken to accomplish often presented great difficulties, and his efforts to overcome them amounted to a succession of individual struggles, sometimes rising almost to the heroic. In one case we find the self-born engineer to be a London goldsmith, like Myddelton; in another, a retired sea-captain, like Perry; a wheelwright, like Brindley; an attorney's clerk, like Smeaton; a mathematical instrument maker, like Watt; a millwright, like Rennie; a working mason, like Telford; a slater, like Clement; or an engine brakesman, like Stephenson.

These men were strong-minded, resolute, and ingenious, and were impelled to their special pursuits by the force of their constructive instincts. In most cases they had to make for themselves a way; for there were none to point out the road, which, until they entered upon their undertakings, had for the most part been untravelled. To our mind, there is almost a dramatic interest in their noble efforts, their defeats, their triumphs; and their constant rise, in spite of manifold obstructions and difficulties, from obscurity to fame.

It will scarcely be supposed that men so uneducated as these, had anything to do with the making of the history of England. History is principally monopolised by the deeds and misdeeds of kings, warriors, and statesmen; and there is little room left for the engineer or the mechanic. But Peace has its battles and its victories as well as War; the results being much more beneficent in the one case than they are in the other.

Our engineers may be regarded in some measure as the makers of modern civilisation. The problems of political history cannot properly be interpreted without reference to the people themselves—how they lived and how they worked, and what they did to promote the civilisation of the nation to which they belonged. Hence English engineers are not unworthy to be considered in the history of their country. For what were England without its roads, its bridges, its canals, its docks, and its harbours. What were it without its tools, its machinery, its steam-engine, its steam-ships, and its locomotive. Are not the men who have made the motive power of the country, and immensely increased its productive strength, the men above all others who have tended to make the country what it is?

The last letter that the author received from Mr. Cobden, contained the following passage :—

" I venture the prediction that not only an enduring but an increasing renown will attach to the memories of those 'Captains of Industry' whose biographies you have recorded; for it cannot be doubted that each succeeding generation will hold in higher estimation those discoveries in physical science to which mankind must attribute henceforth so largely its progress and improvement."

The object of this work is to give an account of some of the principal men by whom this nation has been made so great and prosperous as it is—the men by whose skill and industry large tracts of fertile land have been won from the sea, the bog, and the fen, and made available for human habitation and sustenance; who, by their industry, skill, and genius, have made England the busiest of workshops; who have rendered the country accessible in all directions by roads, bridges, canals, and railways; and who have built lighthouses, breakwaters, docks, and harbours, for the protection and accommodation of our vast home and foreign commerce.

When the author first contemplated writing these memoirs, he was dissuaded rather than encouraged. Robert Stephenson did not think that there was anything in his father's Life to attract the attention of the reading public. " The subject," he said, " possessed no literary interest. What do people care about the building of bridges, the excavation of tunnels, and the making of roads and railways? And then there was the 'Life of Telford,' which had been published at a great expense, and had fallen still-born from the press."

The author, however, entertained a different opinion. He had lived amongst engineering men for many years. He had himself known George Stephenson, and had heard many interesting things about him from his professional associates; and he eventually resolved, with the assistance of Robert Stephenson, to proceed with the preparation of his Life. It was published in 1857, and was received by the public with much favour.

The other lives were proceeded with subsequently;

according as time and opportunity permitted. The original work was published at a high price, in consequence of the great expense of the illustrations. But as the edition is now exhausted, the author has been induced to republish the work in a complete form at a reduced price, in the hope that it may thus obtain a more general circulation In some respects it has been considerably enlarged, and in others compressed; but no essential feature of the original narrative will be found omitted.

CONTENTS.

CHAPTER III.

SIR HUGH MYDDELTON — CUTTING OF THE NEW RIVER.

CHAPTER IV.

SIR HUGH MYDDELTON (continued) — HIS OTHER ENGINEERING AND MINING WORKS — HIS DEATH.

CHAPTER V.

CAPTAIN PERRY — STOPPAGE OF DAGENHAM BREACH.

CHAPTER VI.

THE BEGINNING OF CANAL NAVIGATION.

LIFE OF JAMES BRINDLEY.

CHAPTER I.

BRINDLEY'S EARLY YEARS.

CHAPTER II.

BRINDLEY AS MASTER WHEELWRIGHT AND MILLWRIGHT.

CHAPTER III.

THE DUKE OF BRIDGEWATER — BRINDLEY EMPLOYED AS ENGINEER OF HIS CANAL.

CHAPTER IV.

EXTENSION OF THE DUKE'S CANAL TO THE MERSEY.

CHAPTER V.

THE DUKE'S DIFFICULTIES — COMPLETION OF THE CANAL — GROWTH OF MANCHESTER.

CHAPTER VI.

Brindley constructs the Grand Trunk Canal.

CHAPTER VII.

Brindley's last Canal — His Death and Character.

APPENDIX.

PIERRE-PAUL RIQUET, CONSTRUCTOR OF THE GRAND CANAL OF LANGUEDOC.

LIST OF ILLUSTRATIONS.

JAMES BRINDLEY AND THE
EARLY ENGINEERS.

CHAPTER I.

INTRODUCTORY.

IT has taken the labour and the skill of many generations
of men to make England the country that it now is; to
reclaim and subdue its lands for purposes of agriculture, to
build its towns and supply them with water, to render it
easily accessible by means of roads, bridges, canals, and
railways, and to construct lighthouses, breakwaters, docks,
and harbours for the protection and accommodation of its
commerce. Those great works have been the result of the
continuous industry of the nation, and the men who have
designed and executed them are entitled to be regarded in
a great measure as the founders of modern England.

Engineering, like architecture, strikingly marks the
several stages which have occurred in the development of
society, and throws much curious light upon history. The
ancient British encampment, of which many specimens are
still to be found on the summits of hills, with occasional
indications of human dwellings within them in the circular
hollows or pits over which huts once stood,—the feudal castle
perched upon its all but inaccessible rock, provided with
drawbridge and portcullis to secure its occupants against
sudden assault,—then the moated dwelling, situated in the
midst of the champaign country, indicating a growing,
though as yet but half-hearted confidence in the loyalty of

neighbours,—and, lastly, the modern mansion, with its drawing-room windows opening level with the sward of the adjacent country,—all these are not more striking indications of social progress at the different stages in our history, than the reclamation and cultivation of lands won from the sea, the making of roads and building of bridges, the supplying of towns with water, and the construction of canals and railroads for the ready conveyance of persons and merchandise throughout the empire.

In England, as in all countries, men began with making provision for food and shelter. The valleys and low-lying grounds being mostly covered with dense forests, the naturally cleared high lands, where timber would not grow, were doubtless occupied by the first settlers. Tillage was not as yet understood nor practised; the people subsisted by hunting, or upon their herds of cattle, which found ample grazing among the hills of Dartmoor, and on the grassy downs of Wiltshire and Sussex. Numerous remains or traces of ancient dwellings have been found in those districts, as at Bowhill in Sussex, along the skirts of Dartmoor where the hills slope down to the watercourses, and on the Wiltshire downs, where Old Sarum, Stonehenge and Avebury, mark the earliest and most flourishing of the British settlements.

The art of reclaiming, embanking, and draining land, is supposed to have been introduced by men from Belgium and Friesland, who early landed in great numbers along the south-eastern coasts, and made good their footing by the power of numbers, as well as probably by their superior civilization. The lands from which they came had been won by skill and industry from the sea and from the fen; and when they swarmed over into England, they brought their arts with them. The early settlement of Britain by the races which at present occupy it, is usually spoken of as a series of invasions and conquests; but it is probable that it was for the most part effected by a system of colonization, such as is going forward at this day in America, Australia, and New Zealand; and that the immigrants from Friesland, Belgium, and Jutland, secured their settlement by the spade far more than by the sword. Wherever the new men came, they settled themselves down on their several bits of land, which became their holdings; and they bent their backs over the stubborn soil, watering it with their sweat; and delved, and drained, and cultivated it, until it became fruitful. They also spread themselves over the richer arable lands of the interior, the older population receding before them to the hunting and pastoral grounds of the north and west. Thus the men of Teutonic race gradually occupied the whole of the reclaimable land, and became dominant, as is shown by the dominancy of their language, until they were stopped by the hills of Cumberland, of Wales, and of Cornwall. The same process seems to have gone on in the arable districts of Scotland, into which a swarm of colonists from Northumberland poured in the reign of David I., and quietly settled upon the soil, which they proceeded to cultivate. It is a remarkable confirmation of this view of the early settlement of the country by its present races, that the modern English language extends over the whole of the arable land of England and Scotland, and the Celtic tongue only begins where the plough ends.

One of the most extensive districts along the English coast, lying the nearest to the country from which the continental immigrants first landed, was the tract of Romney Marsh, containing about 60,000 acres of land along the south coast of Kent. The reclamation of this tract is supposed to be due to the Frisians. English history does not reach so far back as the period at which Romney Marsh was first reclaimed, but doubtless the work is one of great antiquity. The district is about fourteen miles long and eight broad, divided into Romney Marsh, Wallend Marsh, Denge Marsh, and Guildford Marsh. The tract is a dead, uniform level, extending from Hythe, in Kent, westward to Winchelsea, in Sussex; and it is to this day held from the

sea by a continuous wall or bank, on the solidity of which
the preservation of the district depends, the surface of the
marsh being under the level of the sea at the highest tides.
The following descriptive view of the marsh, taken from
the high ground above the ancient Roman fortress of
Portus Limanis, near the more modern but still ancient
castle of Lymne, will give an idea of the extent and
geographical relations of the district.

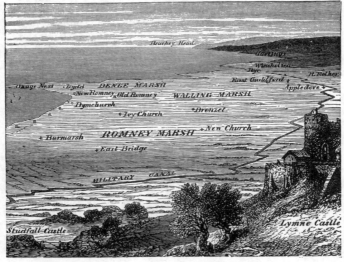

Descriptive View of Romney Marsh, from Lymne Castle.

The tract is so isolated, that the marshmen say the world
is divided into Europe, Asia, Africa, America, and Romney
Marsh. It contains few or no trees, its principal divisions
being formed by dykes and watercourses. It is thinly
peopled, but abounds in cattle and sheep of a peculiarly
hardy breed, which are a source of considerable wealth to the
marshmen; and it affords sufficient grazing for more than
half a million of sheep, besides numerous herds of cattle.

The first portion of the district reclaimed was an island, on which the town of Old Romney now stands; and embankments were extended southward as far as New Romney, where an accumulation of beach took place, forming a natural barrier against further encroachments of the sea at that point. The old town of Lydd originally stood upon another island, as did Ivychurch, Old Winchelsea, and Guildford; the sea sweeping round them and rising far inland at every tide. Burmarsh, and the districts thereabout, were reclaimed at a more recent period; and by degrees the islands disappeared, the sea was shut out, and the whole became firm land. Large additions were made to it from time to time by the deposits of shingle along the coast, which left several towns, formerly important seaports, stranded upon the beach far inland. Thus the ancient Roman port at Lymne, past which the Limen or Rother is supposed originally to have flowed, is left high and dry more than three miles from the sea, and sheep now graze where formerly the galleys of the Romans rode. West Hythe, one of the Cinque Ports, originally the port for Boulogne, is silted up by the wide extent of shingle used by the modern School of Musketry as their practising-ground. Old Romney, past which the Rother afterwards flowed, was one of the ancient ports of the district, but it is now about two miles from the sea. The marshmen followed up the receding waters, and founded the town of New Romney, which also became a Cinque Port; but a storm that occurred in the reign of Edward I. so blocked up the Rother with shingle, at the same time breaching the wall, that the river took a new course, and flowed thenceforward by Rye into the sea; and the port of New Romney became lost. The point of Dungeness, running almost due south, gains accumulations of shingle so rapidly from the sea, that it is said to have extended more than a mile seaward within the memory of persons living. Rye was founded on the ruins of the Romneys, and also became a Cinque Port; but notwithstanding the advantage of the river Rother flowing past it, that port also has become nearly silted up, and now stands about two miles from the sea. New Winchelsea, the Portsmouth and Spithead of its day, is left stranded like the rest of the old Cinque Ports, and is now but a village surrounded by the remains of its ancient grandeur. All this ruin, however, wrought by the invasions of the shingle upon the seacoast towns, has only served to increase the area of the rich grazing ground of the marsh, which continues year by year to extend itself seaward.

Another highly important work of the same class was the embankment of the Thames, now the watery highway between the capital of Great Britain and the world. Before human industry had confined the river within its present channel, it was a broad estuary, in many parts between London and Gravesend several miles wide. The higher tides covered Plumstead and Erith Marshes on the south, and Plaistow, East Ham, and Barking Levels on the north; the river meandering in many devious channels at low water, leaving on either side vast expanses of rich mud and ooze. Opposite the City of London, the tides washed

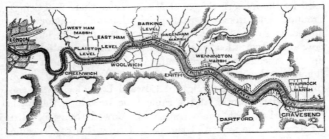

Valley of the Thames (Western Part.)   [From the Ordnance Survey.]

over the ground now covered by Southwark and Lambeth; the district called Marsh still reminding us of its former state, as Bankside informs us of the mode by which it was reclaimed by the banking out of the tidal waters.

A British settlement is supposed to have been formed at

an early period on the high ground on which St. Paul's
Cathedral stands, by reason of its natural defences, being
bounded on the south by the Thames, on the west by the
Fleet, and on the north and east by morasses, Moorfields
Marsh having only been reclaimed within a comparatively
recent period. The natural advantages of the situation
were great, and the City seems to have acquired considerable importance even before the Roman period. The
embanking of the river has been attributed to that indefatigable people; but on this point no evidence exists.
The numerous ancient British camps found in all parts of
the kingdom afford sufficient proof that the early inhabitants of the country possessed a knowledge of the art of
earthwork; and it is not improbable that the same Belgian
tribes who reclaimed Romney Marsh were equally quick to
detect the value for agricultural purposes of the rich allu-
vial lands along the valley of the Thames, and proceeded
accordingly to embank them after the practice of the
country from which they had come. The work was carried
on from one generation to another, as necessity required,

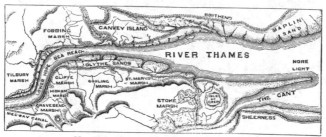

Map of the Valley of the Thames (Eastern Part).
N.B. The dotted line represents the embankments raised along the banks of the river.

until the Thames was confined within its present limits,
the process of embanking serving to deepen the river and
improve it for purposes of navigation, while large tracts of
fertile land were at the same time added to the food-pro-
ducing capacity of the country.

Another of the districts won from the sea, in which a struggle of skill and industry against the power of water, both fresh and salt, has been persistently maintained for centuries, is the extensive low-lying tract of country, situated at the junction of the counties of Lincoln, Huntingdon, Cambridge, and Norfolk, commonly known as the Great Level of the Fens. The area of this district presents almost the dimensions of a province, being from sixty to seventy miles from north to south, and from twenty to thirty miles broad, the high lands of the interior bounding it somewhat in the form of a horse-shoe. It contains about 680,000 acres of the richest land in England, and is as much the product of art as the kingdom of Holland, opposite to which it lies.

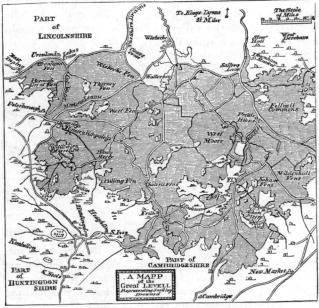

Map of the Fens as they lay drowned. (After Dugdale.)

Not many centuries ago, this vast tract of about two thousand square miles of land was entirely abandoned to the waters, forming an immense estuary of the Wash, into which the rivers Witham, Welland, Glen, Nene, and Ouse discharged the rainfall of the central counties of England. It was an inland sea in winter, and a noxious swamp in summer, the waters expanding in many places into settled seas or meres, swarming with fish and screaming with wild-fowl. The more elevated parts were overgrown with tall reeds, which appeared at a distance like fields of waving corn ; and they were haunted by immense flocks of star-lings, which, when disturbed, would rise in such numbers as almost to darken the air. Into this great dismal swamp the floods descending from the interior were carried, their waters mingling and winding by many devious channels before they reached the sea. They were laden with silt, which became deposited in the basin of the Fens. Thus the river-beds were from time to time choked up, and the intercepted waters forced new channels through the ooze, meandering across the level, and often winding back upon themselves, until at length the surplus waters, through many openings, drained away into the Wash. Hence the numerous abandoned beds of old rivers still traceable amidst the Great Level of the Fens—the old Nene, the old Ouse, and the old Welland. The Ouse, which in past times flowed into the Wash at Wisbeach (or Ouse Beach), now enters at King's Lynn, near which there is another old Ouse. But the probability is that all the rivers flowed into a lake, which existed on the tract known as the Great Bed-ford Level, from thence finding their way, by numerous and frequently shifting channels, into the sea.

Along the shores of the Wash, where the fresh and salt waters met, the tendency to the deposit of silt was the greatest ; and in the course of ages, the land at the outlets of the inland waters became raised above the level of the interior. Accordingly, the first land reclaimed in the dis-trict was the rich fringe of deposited silt lying along the shores of the Wash, now known as Marshland and South Holland. This was effected by the Romans, a hard-working, energetic, and skilful people; of whom the Britons are said to have complained * that they wore out and consumed their hands and bodies in clearing the woods and banking the fens. The bulwarks or causeways which they raised to keep out the sea are still traceable at Po-Dyke in Marshland, and at various points near the old coast-line. On the inland side of the Fens the Romans are supposed to have constructed another great work of drainage, still known as Carr Dyke, extending from the Nene to the Witham. It means Fen Dyke, the fens being still called Carrs in certain parts of Lincoln. This old drain is about sixty feet wide, with a broad, flat bank on each side ; and originally it must have been at least forty miles in extent, winding along under the eastern side of the high land, which extends in an irregular line up the centre of the district from Stamford to Lincoln.

The eastern parts of Marshland and Holland were thus the first lands reclaimed in the Level, and they were available for purposes of agriculture long before any attempts had been made to drain the lands of the interior. Indeed, it is not improbable that the early embankments thrown up along the coast had the effect of increasing the inundations of the lower-lying lands farther west; for, whilst they dammed the salt water out, they also held back the fresh, no provision having been made for improving and deepening the outfalls of the rivers flowing through the Level into the Wash. The Fen lands in winter were thus not only flooded by the rainfall of the Fens themselves, and by the upland waters which flowed from the interior, but also by the daily flux of the tides which drove in from the German Ocean, holding back the fresh waters, and even mixing with them far inland.

The Fens, therefore, continued flooded with water down

---

* Tacitus, 'Life of Agricola.'

to the period of the Middle Ages, when there was water
enough in the Witham to float the ships of the Danish sea
rovers as far inland as Lincoln, where ships' ribs and
timbers have recently been found deep sunk in the bed
of the river. The first reclaimers of the Fen lands seem
to have been the religious recluses, who settled upon the
islands overgrown with reeds and rushes, which rose up
at intervals in the Fen level, and where they formed
their solitary settlements. One of the first of the Fen
islands thus occupied was the Isle of Ely, or Eely—so
called, it is said, because of the abundance and goodness
of the eels caught in the neighbourhood, and in which
rents were paid in early times. It stood solitary amid
the waste of waters, and was literally an island. Ethel-
dreda, afterwards known as St. Audrey, the daughter of
the King of the East Angles, retired thither, secluding
herself from the world and devoting herself to a recluse
life. A nunnery was built, then a town, and the place
became famous in the religious world. The pagan Danes,
however, had no regard for Christian shrines, and a fleet of
their pirate ships, sailing across the Fens, attacked the
island and burnt the nunnery. It was again rebuilt, and a
church sprang up, the fame of which so spread abroad that
Canute, the Danish king, determined to visit it. It is related that as his ships sailed towards the island his soul
rejoiced greatly, and on hearing the chanting of the monks
in the quire wafted across the waters, the king joined in
the singing and ceased not until he had come to land.
Canute more than once sailed across the Fens with his
ships, and the tradition survives that on one occasion, when
passing from Ramsey to Peterborough, the waves were so
boisterous on Whittlesea Mere (now a district of fruitful
cornfields), that he ordered a channel to be cut through the
body of the Fen westward of Whittlesea to Peterborough,
which to this day is called by the name of the "King's
Delph."

The other Fen islands which became the centres of subsequent reclamations were Crowland, Ramsey, Thorney, and Spinney, each the seat of a monastic establishment. The old churchmen, notwithstanding their industry, were, however, only able to bring into cultivation a few detached points, and made very little impression upon the drowned lands of the Great Level. It often happened, indeed, that the steps which they took to drain one spot merely had the effect of sending an increased flood of water upon another, and perhaps diverting in some new direction the water which before had driven a mill, or formed a channel for purposes of navigation. The rivers also were constantly liable to get silted up, and form for themselves new courses; and sometimes, during a high tide, the sea would burst in, and in a single night undo the tedious industry of centuries.

Each suffering locality, acting for itself, did what it could to preserve the land which had been won, and to prevent the recurrence of inundations. Dyke-reeves were appointed along the sea-borders, with a force of shore-labourers at their disposal, to see to the security of the embankments; and fen-wards were constituted inland, over which commissioners were set, for the purpose of keeping open the drains, maintaining the dykes, and preventing destruction of life and property by floods, whether descending into the Fens from the high lands or bursting in upon them from the sea. Where lands became suddenly drowned, the Sheriff was authorised to impress diggers and labourers for raising embankments; and commissioners of sewers were afterwards appointed, with full powers of local action, after the law and usage of Romney Marsh. In one district we find a public order made that every man should plant with willows the bank opposite his portion of land towards the fen, "so as to break off the force of the waves in flood times;" and swine were not to be allowed to go upon the banks unless they were ringed, under a penalty of a penny (equal to a shilling in our money) for every hog found unringed. A still more terrible penalty for neglect is mentioned by Harrison, who says, " Such as having walls or banks near unto the sea,

and do suffer the same to decay (after convenient admoni-
tion), whereby the water entereth and drowneth up the
country, are by a certain ancient custom apprehended, con-
demned, and *staked in the breach*, where they remain for ever
as parcel of the new wall that is to be made upon them, as
I have heard reported."*

The Great Level of the Fens remained in a comparatively
unreclaimed state down even to the end of the sixteenth
century; and constant inundations took place, destroying
the value of the little settlements which had by that time
been won from the watery waste.   It would be difficult to
imagine anything more dismal than the aspect which the
Great Level then presented.   In winter, a sea without
waves; in summer, a dreary mud-swamp.   The atmosphere
was heavy with pestilential vapours, and swarmed with
insects.   The meres and pools were, however, rich in fish and
wild-fowl.   The Welland was noted for sticklebacks, a little
fish about two inches long, which appeared in dense shoals
near Spalding every seventh or eighth year, and used to be
sold during the season at a halfpenny a bushel, for field
manure.   Pike was plentiful near Lincoln : hence the
proverb, "Witham pike, England hath none like."   Fen-
nightingales, or frogs, especially abounded.   The birds-
proper were of all kinds; wild-geese, herons, teal, widgeons,
mallards, grebes, coots, godwits, whimbrels, knots, dottrels,
yelpers, ruffs, and reeves, some of which have long since
disappeared from England.   Mallards were so plentiful
that 3000 of them, with other birds in addition, have been
known to be taken at one draught.   Round the borders of
the Fens there lived a thin and haggard population of "Fen-
slodgers," called "yellow-bellies" in the inland counties,
who derived a precarious subsistence from fowling and
fishing.   They were described by writers of the time as "a
rude and almost barbarous sort of lazy and beggarly people."
Disease always hung over the district, ready to pounce upon

---

* Harrison's Preface to Holinshed's Chronicle, i. 313.

the half-starved fenmen. Camden spoke of the country between Lincoln and Cambridge as " a vast morass, inhabited by fenmen, a kind of people, according to the nature of the place where they dwell, who, walking high upon stilts, apply their minds to grazing, fishing, or fowling." The proverb of " Cambridgeshire camels " doubtless originated in this old practice of stilt-walking in the Fens ; the fenmen, like the inhabitants of the Landes, mounting upon high stilts to spy out their flocks across the dead level. But the flocks of the fenmen consisted principally of geese, which were called the " fenmen's treasure ;" the fenman's dowry being " three-score geese and a pelt " or sheep-skin used as an outer garment. The geese throve where nothing else could exist, being equally proof against rheumatism and ague, though lodging with the natives in their sleeping-places. Even of this poor property, however, the slodgers were liable at any time to be stripped by sudden inundations.

In the oldest reclaimed district of Holland, containing many old village churches, the inhabitants, in wet seasons, were under the necessity of rowing to church in their boats. In the other less reclaimed parts of the Fens the inhabitants were much worse off. " In the winter time," said Dugdale, " when the ice is only strong enough to hinder the passage of boats, and yet not able to bear a man, the inhabitants upon the hards and banks within the Fens can have no help of food, nor comfort for body or soul ; no woman aid in her travail, no means to baptize a child or partake of the Communion, nor supply of any necessity saving what these poor desolate places do afford. And what expectation of health can there be to the bodies of men, where there is no element good ? the air being for the most part cloudy, gross, and full of rotten harrs ; the water putrid and muddy, yea, full of loathsome vermin ; the earth spungy and boggy, and the fire noisome by the stink of smoaky hassocks."

The wet character of the soil at Ely may be inferred from the circumstance that the chief crop grown in the neighbourhood was willows; and it was a common saying there, that "the profit of willows will buy the owner a horse before that by any other crop he can pay for his saddle." There was so much water constantly lying above Ely, that in olden times the Bishop of Ely was accustomed to go in his boat to Cambridge. When the outfalls of the Ouse became choked up by neglect, the surrounding districts were subject to severe inundations; and after a heavy fall of rain, or after a thaw in winter, when the river swelled suddenly, the alarm spread abroad, "the bailiff of Bedford is coming!" the Ouse passing by that town. But there was even a more terrible visitor than the bailiff of Bedford; for when a man was stricken down by the ague, it was said of him, "he is arrested by the bailiff of Marsh-land;" this disease extensively prevailing all over the district when the poisoned air of the marshes began to work.

The great perils which constantly threatened the district at length compelled the attention of the legislature. In 1607, shortly after the accession of James I., a series of destructive floods burst in the embankments along the east coast, and swept over farms, homesteads, and villages, drowning large numbers of people and cattle. When the King was informed of the great calamity which had befallen the inhabitants of the Fens, principally through the decay of the old works of drainage and embankment, he is said to have made the right royal declaration, that "for the honour of his kingdom, he would not any longer suffer these countries to be abandoned to the will of the waters, nor to let them lie waste and unprofitable; and that if no one else would undertake their drainage, he himself would become their undertaker." A Commission was appointed to inquire into the extent of the evil, from which it appeared that there were not less than 317,242 acres of land lying outside the then dykes which required drainage and protection. A bill was brought into Parliament to enable rates to be levied for the drainage of this land, but it was summarily rejected. Two years later. a "little bill," for draining

6000 acres in Waldersea County, was passed — the first
district Act for Fen drainage that received the sanction of
Parliament. The King then called Chief-Justice Popham
to his aid, and sent him down to the Fens to undertake a
portion of the work; and he induced a company of Londoners to undertake another portion, the adventurers
receiving two-thirds of the reclaimed lands as a recompense. "Popham's Eau," and "The Londoners' Lode,"
still mark the scene of their operations. The works, however, did not prove very successful, not having been carried
out with sufficient practical knowledge on the part of the
adventurers, nor after any well-devised plan. There were
loud calls for some skilled undertaker or engineer (though
the latter word was not then in use) to stay the mischief,
reclaim the drowned lands, and save the industrious settlers
in the Fens from total ruin. But no English engineer was
to be found ready to enter upon so large an undertaking ;
and in his dilemma the King called to his aid one Cornelius Vermuyden, a Dutch engineer, a man well skilled
in works of embanking and draining.

The necessity for employing a foreign engineer to undertake so great a national work is sufficiently explained by
the circumstance that England was then very backward in
all enterprises of this sort. We had not yet begun that
career of industrial skill in which we have since achieved
so many triumphs, but were content to rely mainly upon
the assistance of foreigners. Holland and Flanders supplied us with our best mechanics and engineers. Not
only did Vermuyden prepare the plans and superintend
the execution of the Great Level drainage, but the works
were principally executed by Flemish workmen. Many
other foreign "adventurers" as they were called, besides
Vermuyden, carried out extensive works of reclamation
and embankment of waste lands in England. Thus a
Fleming named Freeston reclaimed the extensive marsh
near Wells in Norfolk; Joas Croppenburgh and his company of Dutch workmen reclaimed and embanked Canvey

Island near the mouth of the Thames; Cornelius Vander-welt, another Dutchman, enclosed Wapping Marsh by means of a high bank, along which a road was made, called "High Street" to this day; while two Italians, named Acontius and Castilione, reclaimed the Combe and East Greenwich marshes on the south bank of the river.

We also relied very much on foreigners for our harbour engineering. Thus, when a new haven was required at Yarmouth, Joas Johnson, the Dutchman, was employed to plan and construct it. When a serious breach occurred in the banks of the Witham at Boston, Mathew Hakes was sent for from Gravelines, in Flanders, to repair it; and he brought with him not only the mechanics, but the manufac-tured iron required for the work. In like manner, any unusual kind of machinery was imported from Holland or Flanders ready made. When an engine was needed to pump water from the Thames for the supply of London, Peter Morice, the Dutchman, brought one from Holland, together with the necessary workmen.

England was in former times regarded principally as a magazine for the supply of raw materials, which were carried away in foreign ships, and returned to us worked up by foreign artisans. We grew wool for Flanders, as India, America, and Egypt grow cotton for England now. Even the wool manufactured at home was sent to the Low Countries to be dyed. Our fisheries were so unproductive, that the English markets were supplied by the Dutch, who sold us the herrings caught in our own seas, off our own shores. Our best ships were built for us by Danes and Genoese; and when any skilled sailors' work was wanted, foreigners were employed. Thus, when the "Mary Rose" sank at Spithead in 1545, Peter de Andreas, the Venetian, with his ship carpenter and three Italian sailors, were employed to raise her, sixty English mariners being appointed to attend upon them merely as labourers.

In short, we depended for our engineering, even more than we did for our pictures and our music, upon foreigners.

Nearly all the continental nations had a long start of us in
art, in science, in mechanics, in navigation, and in engineering. At a time when Holland had completed its magnificent system of water communication, and when France,
Germany, and even Russia had opened up important lines
of inland navigation, England had not cut a single canal,
whilst our roads were about the worst in Europe.

CHAPTER II.

Sir Cornelius Vermuyden — Drainage of the Fens.

Cornelius Vermuyden, the Dutch engineer, was invited over to England about the year 1621, to stem a breach in the Thames embankment near Dagenham, which had been burst through by the tide. He was a person of good birth and education, and was born at St. Martin's Dyke, in the island of Tholen, in Zealand. He had been trained as an engineer, and having been brought up in a district where embanking was studied as a profession, and gave employment to a large number of persons, he was familiar with the most approved methods of protecting land against the encroachments of the sea. He was so successful in his operations at Dagenham, that when it was found necessary to drain the Royal park at Windsor, he was employed to conduct the work ; and he thus became known to the king, who shortly after employed him in the drainage of Hatfield Level, then a royal chase on the borders of Yorkshire.

The extensive district of Axholme, of which Hatfield Chase formed only a part, resembled the Great Level of the Fens in many respects, being a large fresh-water bay formed by the confluence of the rivers Don, Went, Ouse, and Trent, which brought down into the Humber almost the entire rainfall of Yorkshire, Derbyshire, Nottingham, and North Lincoln, and into which the sea also washed. The uplands of Yorkshire bounded this watery tract on the west, and those of Lincolnshire on the east. Rising up about midway between them was a single hill, or rather elevated ground, formerly an island, and still known as the

Isle of Axholme. There was a ferry between Sandtoft and
that island in times not very remote, and the farmers of
Axholme were accustomed to attend market at Doncaster in
their boats, though the bottom of the sea over which they
then rowed is now amongst the most productive corn-land
in England. The waters extended to Hatfield, which lies
along the Yorkshire edge of the level on the west; and it is
recorded in the ecclesiastical history of that place that a
company of mourners, with the corpse they carried, were
once lost when proceeding by boat from Thorne to Hatfield.
When Leland visited the county in 1607, he went by boat
from Thorne to Tudworth, over what at this day is rich
ploughed land. The district was marked by numerous
merestones, and many fisheries are still traceable in local
history as having existed at places now far inland.

The Isle of Axholme was in former times a stronghold of
the Mowbrays, being unapproachable save by water. In the
reign of Henry II., when Lord Mowbray held it against the
King, it was taken by the Lincolnshire men, who attacked
it in boats; and, down to the reign of James I., the only
green spot which rose above the wide waste of waters was
this solitary isle. Before that monarch's time the south-
eastern part of the county of York, from Conisborough Castle
to the sea, belonged for the most part to the crown; but
one estate after another was alienated, until at length, when
James succeeded to the throne of England, there only
remained the manor of Hatfield, which, watery though it
was, continued to be dignified with the appellation of a
Royal Chase. There was, however, plenty of deer in the
neighbourhood, for De la Pryme says that in his time
they were as numerous as sheep on a hill, and that venison
was as abundant as mutton in a poor man's kitchen.*
But the principal sport which Hatfield furnished was
in the waters and meres adjacent to the old timber
manor-house. Prince Henry, the King's eldest son, on the

---

* De la Pryme, 'History of the Level of Hatfield Chase.'

occasion of a journey to York, rested at Hatfield on his way, and had a day's sport in the Royal Chase, which is thus described by De la Pryme :—" The prince and his retinue all embarked themselves in almost a hundred boats that were provided there ready, and having frightened some five hundred deer out of the woods, grounds, and closes adjoining, which had been drawn there the night before, they all, as they were commonly wont, took to the water, and this little royal navy pursuing them, soon drove them into that lower part of the level, called Thorne Mere, and there, being up to their very necks in water, their horned heads raised themselves so as almost to represent a little wood. Here being encompassed about with the little fleet, some ventured amongst them, and feeling such and such as were fattest, they either immediately cut their throats, or else tying a strong long rope to their heads, drew them to land and killed them."

Such was the last battue in the Royal Chase of Hatfield. Shortly after, King James brought the subject of the drainage of the tract under the notice of Cornelius Vermuyden, who, on inspecting it, declared the project to be quite practicable. The level of the Chase contained about 70,000 acres, the waters of which, like those of the Fens, found their way to the sea through many changing channels. Various attempts had been made to diminish the flooding of the lands. In the fourteenth century several deep trenches were dug, to let off the water, but they probably admitted as much as they allowed to escape, and the drowning continued. Commissioners were appointed, but they did nothing. The country was too poor, and the people too unskilled, to undertake so expensive and laborious an enterprise as the effectual drainage of so large a tract.

A local jury was summoned by the King to consider the question, but they broke up, after expressing their opinion of the utter impracticability of carrying out any effective plan for the withdrawal of the waters. Vermuyden, however, declared that he would undertake

and bind himself to do that which the jury had pronounced
to be impossible. The Dutch had certainly been successful
beyond all other nations in projects of the same kind. No
people had fought against water so boldly, so perseveringly,
and so successfully. They had made their own land out of
the mud of the rest of Europe, and, being rich and prosperous, were ready to enter upon similar enterprises in
other countries. On the death of James I., his successor
confirmed the preliminary arrangement which had been
made with Vermuyden, with a view to the drainage of
Hatfield Manor; and on the 24th of May, 1626, after a
good deal of negotiation as to terms, articles were drawn up
and signed between the Crown and Vermuyden, by which
the latter undertook to reclaim the drowned lands, and
make them fit for tillage and pasturage. It was a condition
of the contract that Vermuyden and his partners in the
adventure were to have granted to them one entire third of
the lands so recovered from the waters.

Vermuyden was a bold and enterprising man, full of
energy and resources. He also seems to have possessed the
confidence of capitalists in his own country, for we find him
shortly after proceeding to Amsterdam to raise the requisite
money, of which England was then so deficient; and a company was formed composed almost entirely of Dutchmen, for
the purpose of carrying out the necessary works of reclamation. Amongst those early speculators in English drainage
we find the names of the Valkenburgh family, the Van
Peenens, the Vernatti, Andrew Boccard, and John Corsellis.
Of the whole number of shareholders amongst whom the
lands were ultimately divided, the only names of English
sound are those of Sir James Cambell, Knight, and Sir
John Ogle, Knight, who were about the smallest of the
participants.

Several of the Dutch capitalists came over in person to
look after their respective interests in the concern, and
Vermuyden proceeded to bring together from all quarters
a large number of workmen, mostly Dutch and Flemish.

It so happened that there were then settled in England numerous foreign labourers — Dutchmen who had been brought from Holland to embank the lands at Dagenham and Canvey Island on the Thames, and others who had been driven from their own countries by religious persecution—French Protestants from Picardy, and Walloons from Flanders.   The countries in which those people had been born and bred resembled in many respects the marsh and fen districts of England, and they were practically familiar with the reclamation of such lands, the digging of drains, the raising of embankments, and the cultivation of marshy ground.   Those immigrants had already settled down in large numbers in the eastern counties, and along the borders of the Fens, at Wisbeach, Whittlesea, Thorney, Spalding, and the neighbourhood.*   The poor foreigners readily answered Vermuyden's call, and many of them took service under him at Hatfield Chase, where they set to work with such zeal, and laboured with such diligence, that before the end of the second year the work was so far advanced, that a commission was issued for the survey and division amongst the participants of the reclaimed lands.

The plan of drainage adopted seems to have been, to carry the waters of the Idle by direct channels into the Trent, instead of allowing them to meander at will through the level of the Chase.   Deep drains were cut, through which the water was drawn from the large pools standing near Hatfield and Thorne.   The Don also was blocked out of the level by embankments, and forced through its northern branch, by Turn-bridge, into the river Aire.   But

---

* It has been observed that the buildings in many of the old Fen towns to this day have a Flemish appearance, as the names of many of the inhabitants have evidently a foreign origin.   Those of Descow, Le Plas, Egar, Bruynne, &c., are said to be still common.   Among the settlers in the level of Hatfield was Mathew de la Pryme, who emigrated from Ypres in Flanders during the persecutions of the Duke of Alva.   The Prymes of Cambridge are lineally descended from him.   Tablets to several of the name are still to be found in Hatfield Church.

this last attempt proved a mistake, for the northern channel was found insufficient for the discharge of the waters, and floodings of the old lands about Fishlake, Sykehouse, and

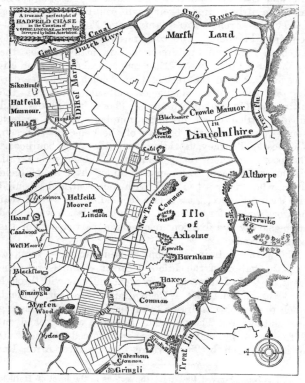

Map of the Level of Hatfield Chase.   [Corrected, after Dugdale.]

Snaith took place; to prevent which, a wide and deep channel, called the Dutch River, was afterwards cut, and the waters of the Don were sent directly into the Ouse, near Goole. This great and unexpected addition to the cost of the undertaking appears to have had a calamitous effect, and brought distress and ruin on many who had engaged in it. The people who dwelt on the northern branch of the Don complained loudly of the adventurers, who were denounced as foreigners and marauders; and they were not satisfied with mere outcry, but took the law into their own hands; broke down the embankments, assaulted the Flemish workmen, and several persons lost their lives in the course of the riots which ensued.*

Vermuyden did what he could to satisfy the inhabitants. He employed large numbers of native workmen, at considerably higher wages than had before been paid; and he strenuously exerted himself to relieve those who had suffered from the changes he had effected, so far as could be done without incurring a ruinous expense.† Dugdale relates that there could be no question about the great benefits which the execution of the drainage works conferred upon the labouring population; for whereas, before the reclamation, the country round about had been "full of wandering

---

* R. Ansbie writes the Duke of Buckingham from Tickill Castle, under date the 21st August, 1628, as follows :—" What has happened betwixt Mr. Vermuyden's friends and workmen and the people of the Isle of Axholme these inclosed will give a taste. Great riots have been committed by the people, and a man killed by the Dutch party, the killing of whom is conceived to be murder in all who gave direction for them to go armed that day. These outrages will produce good effects. They will procure conformity in the people, and enforce Vermuyden to sue for favour at the Duke's hands,—if not for himself, for divers of his friends, especially for Mr. Saines, a Dutchman, who has an adventure of 13,000l. in this work. Upon examination of the rest of Vermuyden's people, thinks it will appear that he gave them orders to go armed."—'State Papers,' vol. cxiii. 38.

† F. Vernatti, one of the Dutch capitalists who had contributed largely towards the cost of the works, writes to Monsieur St. Gillis, in October, 1628 :—" The absence of Mr. Vermuyden, and the great interest the writer takes in the business of embankment at Haxey, has led him to engage in it with eye and hand. The mutinous people have not only desisted from their threats, but now give their work to complete the dyke, which they have fifty times destroyed and thrown into the river. A royal proclamation made by a serjeant-at-arms in their village, accompanied by the sheriff and other officials, with fifty horsemen, and an exhortation mingled with threats of fire and vengeance, have produced this result." —'State Papers,' vol. cxix. 73.

beggars," these had now entirely disappeared, and there
was abundant employment for all who would work, at good
wages. An immense tract of rich land had been completely
recovered from the waters, but it could only be made
valuable and productive after long and diligent cultivation.

Vermuyden was throughout well supported by the
Crown, and on the 6th of January, 1629, he received the
honour of knighthood at the hands of Charles I., in recognition of the skill and energy which he had displayed in adding so large a tract to the cultivable lands of England. In
the same year he took a grant from the Crown of the whole
of the reclaimed lands in the manor of Hatfield, amounting to about 24,500 acres, agreeing to pay the Crown the
sum of 16,080l., an annual rent of 193l. 3s. 5½d., one red rose
ancient rent, and an improved rent of 425l. from Christmas,
1630.* Power was also granted him to erect one or more
chapels wherein the Dutch and Flemish settlers might
worship in their own language. They built houses, farm-
steads, and windmills; intending to settle down peacefully
to cultivate the soil which their labours had won.

It was long, however, before the hostility and jealousy of
the native population could be appeased. The idea of
foreigners settling as colonists upon lands over which,
though mere waste and swamp, their forefathers had
enjoyed rights of common, was especially distasteful to
them, and bred bitterness in many hearts. The dispos-
sessed fenmen had numerous sympathisers among the rest
of the population. Thus, on one occasion, we find the
Privy Council sending down a warrant to all Postmasters
to furnish Sir Cornelius Vermuyden with horses and a
guide to enable him to ride post from London to Boston,
and from thence to Hatfield.† But at Royston " Edward
Whitehead, the constable, in the absence of the postmaster,
refused to provide horses, and on being told he should

---

* 'State Papers,' vol. cxlvii. 21. | Whitehall, May 12th, 1630.—'State
† Warrant of Council, dated | Papers,' vol. clxvi. 56.

answer for his neglect, replied, 'Tush! do your worst: you shall have none of my horses in spite of your teeth.'"[*] Complaints were made to the Council of the injury done to the surrounding districts by the drainage works; and an inquisition was held on the subject before the Earls of Clare and Newcastle, and Sir Gervase Clifton. Vermuyden was heard in defence, and a decision was given in his favour; but he seems to have acted with precipitancy in taking out subpœnas against many of the old inhabitants for damage said to have been done to him and his agents. Several persons were apprehended and confined in York gaol, and the feeling of bitterness between the native population and the Dutch settlers grew more intense from day to day. Lord Wentworth, President of the North, at length interfered; and after surveying the lands, he ordered that all suits should cease. Vermuyden was also directed to assign to the tenants certain tracts of moor and marsh ground, to be enjoyed by them in common. He attempted to evade the decision, holding it to be unjust; but the Lord President was too powerful for him, and feeling that further opposition was of little use, he resolved to withdraw from the undertaking, which he did accordingly; first conveying his lands to trustees, and afterwards disposing of his interest in them altogether.[†]

The necessary steps were then taken to relieve the old

---

[*] Affidavit of George Johnson, servant of Sir Cornelius Vermuyden. —'State Papers,' vol. clxx. 17.

[†] The Dutch settlers lived for the most part in single houses, dispersed through the newly-recovered country. A house built by Vermuyden remains. It was chiefly of timber, and what is called *stud-bound*. It was built round a quadrangular court. The eastern front was the dwelling-house. The other three sides were stables and barns. Another good house was built by Mathew Valkenburgh, on the Middle

Ing, near the Don, which afterwards became the property of the Boynton family. Sir Philibert Vernatti and the two De Witts erected theirs near the Idle. A chapel for the settlers was also erected at Sandtoft, in which the various ordinances of religion were performed, and the public service was read alternately in the Dutch and French languages.—The Rev. Joseph Hunter's 'History and Topography of the Deanery of Doncaster,' 1828, vol. i. 165-6.

lands which had been flooded, by the cutting of the Dutch
River at a heavy expense. Great difficulty was experienced
in raising the requisite funds; the Dutch capitalists now
holding their hand, or transferring their interest to other
proprietors, at a serious depreciation in the value of their
shares. The Dutch River was, however, at length cut, and
all reasonable ground of complaint so far as respected the
lands along the North Don was removed. For some years
the new settlers cultivated their lands in peace; when suddenly they were reduced to the greatest distress, through
the troubles arising out of the wars of the Commonwealth.

In 1642 a committee sat at Lincoln to watch over the
interests of the Parliament in that county. The Yorkshire
royalists were very active on the other side of the Don, and
the rumour went abroad that Sir Ralph Humby was about
to march into the Isle of Axholme with his forces. To
prevent this, the committee at Lincoln gave orders to break
the dykes, and pull up the flood-gates at Snow-sewer and
Millerton-sluice. Thus in one night the results of many
years' labour were undone, and the greater part of the level
again lay under water. The damage inflicted on the
Hatfield settlers in that one night was estimated at not less
than 20,000l. The people who broke the dykes were, no
doubt, glad to have the opportunity of taking their full
revenge upon the foreigners for robbing them of their
commons. They levelled the Dutchmen's houses, destroyed
their growing corn, and broke down their fences; and,
when some of them tried to stop the destruction of the
sluices at Snow-sewer, the rioters stood by with loaded
guns, and swore they would stay until the whole levels
were drowned again, and the foreigners forced to swim
away like ducks.

After the mischief had been done, the commoners set up
their claims as participants in the lands which had not been
drowned, from which the foreigners had been driven. In
this they were countenanced by Colonel Lilburne, who,
with a force of Parliamentarians, occupied Sandtoft, driving

the Protestant minister out of his house, and stabling their horses in his chapel. A bargain was actually made between the Colonel and the commoners, by which 2000 acres of Epworth Common were to be assigned to him, on condition of their right being established as to the remainder, while he undertook to hold them harmless in respect of the cruelties which they had perpetrated on the poor settlers of the level. When the injured parties attempted to obtain redress by law, Lilburne, by his influence with the Parliament, the army, and the magistrates, parried their efforts for eleven years.* He was, however, eventually compelled to disgorge; and though the original settlers at length got a decree of the Council of State in their favour, and those of them who survived were again permitted to occupy their holdings, the nature of the case rendered it impossible that they should receive any adequate redress for their losses and sufferings.†

In the mean time Sir Cornelius Vermuyden had not been idle. He was as eagerly speculative as ever. Before he parted with his interest in the reclaimed lands at Hatfield, he was endeavouring to set on foot his scheme for the reclamation of the drowned lands in the Cambridge Fens; for we find the Earl of Bedford, in July, 1630, writing to Sir Harry Vane, recommending him to join Sir Cornelius and himself in the enterprise. Before the end of the year Vermuyden entered into a contract with the Crown for the purchase of Malvern Chase, in the county of Worcester, for

---

* Colonel Lilburne attempted an ineffectual defence of himself in the tract entitled 'The Case of the Tenants of the Manor of Epworth truly stated by Col. Jno. Lilburne,' Nov. 18th, 1651.

† For a long time after this, indeed, the commoners continued at war with the settlers, and both were perpetually resorting to the law—of the courts as well as of the strong hand. One Reading, a counsellor, was engaged to defend the rights of the drainers or participants, but his office proved a very dangerous one. The fen-men regarded him as an enemy, and repeatedly endeavoured to destroy him. Once they had nearly burned him and his family in their beds. Reading died in 1716, at a hundred years old, fifty of which he had passed in constant danger of personal violence, having fought "thirty-one set battles" with the fen-men in defence of the drainers' rights.

the sum of 5000l., which he forthwith proceeded to reclaim and enclose. Shortly after he took a grant of 4000 acres of waste land on Sedgemoor, with the same object, for which he paid 12,000l. Then in 1631 we find him, in conjunction with Sir Robert Heath, taking a lease for thirty years of the Dovegang lead-mine, near Wirksworth, reckoned the best in the county of Derby. But from this point he seems to have become involved in a series of lawsuits, from which he never altogether shook himself free. His connection with the Hatfield estates got him into legal, if not pecuniary difficulties, and he appears for some time to have suffered imprisonment. He was also harassed by the disappointed Dutch capitalists at the Hague and Amsterdam, who had suffered heavy losses by their investments at Hatfield, and took legal proceedings against him. He had no sooner, however, emerged from confinement than we find him fully occupied with his new and grand project for the drainage of the Great Level of the Fens.

The outfalls of the numerous rivers flowing through the Fen Level having become neglected, the waters were everywhere regaining their old dominion. Districts which had been partially reclaimed were again becoming drowned, and even the older settled farms and villages situated upon the islands of the Fens were threatened with like ruin. The Commissioners of Sewers at Huntingdon attempted to raise funds for improving the drainage by levying a tax of six shillings an acre upon all marsh and fen lands, but not a shilling of the tax was collected. This measure having failed, the Commissioners of Sewers of Norfolk, at a session held at King's Lynn, in 1629, determined to call to their aid Sir Cornelius Vermuyden. At an interview to which he was invited, he offered to find the requisite funds to undertake the drainage of the Level, and to carry out the works after the plans submitted by him, on condition that 95,000 acres of the reclaimed lands were granted to him as a recompense. A contract was entered into on those terms; but so great an outcry was immediately raised against such an arrangement being made with a foreigner, that it was abrogated before many months had passed.

Then it was that Francis, Earl of Bedford, the owner of many of the old church-lands in the Fens, was induced to take the place of Vermuyden, and become chief undertaker in the drainage of the extensive tract of fen country now so well known as the Great Bedford Level. Several of the adjoining landowners entered into the project with the Earl, contributing sums towards the work, in return for which a proportionate acreage of the reclaimed lands was to be allotted to them. The new undertakers, however, could not dispense with the services of Vermuyden. He had, after long study of the district, prepared elaborate plans for its drainage, and, besides, had at his command an organized staff of labourers, mostly Flemings, who were well accustomed to this kind of work. Westerdyke, also a Dutchman, prepared and submitted plans, but Vermuyden's were preferred, and he was accordingly authorised to proceed with the enterprise.

The difficulties encountered in carrying on the works were very great, arising principally from the want of funds. The Earl of Bedford became seriously crippled in his resources; he raised money upon his other property until he could raise no more, while many of the smaller undertakers were completely ruined. Vermuyden meanwhile took energetic measures to provide the requisite means to pay the workmen and prosecute the drainage; until the undertakers became so largely his debtors that they were under the necessity of conveying to him many thousand acres of the reclaimed lands, even before the works were completed, as security for the large sums which he had advanced.

The most important of the new works executed at this stage were as follows;—Bedford River (now known as Old Bedford River), extending from Erith on the Ouse to Salter's Lode on the same river: this cut was 70 feet wide and 21 miles long, and its object was to relieve and take off

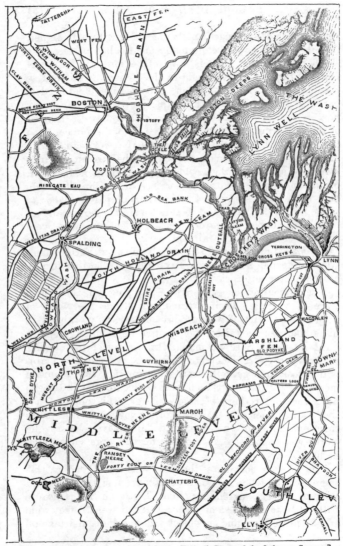

Map of the Fens as drained in 1830. [After Telford's Plan and the Ordnance Survey.]

the high floods of the Ouse.* Bevill's Leam was another extensive cut, extending from Whittlesea Mere to Guyhirne, 40 feet wide and 10 miles long; Sam's Cut, from Feltwell to the Ouse, 20 feet wide and 6 miles long; Sandy's Cut, near Ely, 40 feet wide and 2 miles long; Peakirk Drain, 17 feet wide and 10 miles long; with other drains, such as Mildenhall, New South Eau, and Shire Drain. Sluices were also erected at Tydd upon Shire Drain, at Salter's Lode, and at the Horseshoe below Wisbeach, together with a clow,† at Clow's Cross, to keep out the tides; while a strong fresh-water sluice was also provided at the upper end of the Bedford River.

These works were not permitted to proceed without great opposition on the part of the Fen-men, who frequently assembled to fill up the cuts which the labourers had dug, and to pull down the banks which they had constructed. They also abused and maltreated the foreigners when the opportunity offered, and sometimes mobbed them while employed upon the drains, so that in several places they had to work under a guard of armed men. Difficult though it was to deal with the unreclaimed bogs, the unreclaimed "fen-slodgers" were still more impracticable. Although their condition was very miserable, they nevertheless enjoyed a sort of wild liberty amidst the watery wastes, which they were not disposed to give up. Though they might alternately shiver and burn with ague, and become prematurely bowed and twisted with rheumatism, still the Fens were their "native land," such as it was, and their only source of subsistence, precarious though it might be. The Fens were their commons, on which their geese grazed.

---

* We insert the map on the preceding page at this place, although it includes the drainage-works subsequently constructed, in order that the reader may be enabled more readily to follow the history of the various cuts and drains executed in the Fen country from about the middle of the sixteenth century down to about the year 1830.

† A clow is a sluice regulated by being lifted or dropped perpendicularly, like a portcullis. The other sluices open and shut like gates.

They furnished them with food, though the finding thereof was full of adventure and hazard. What cared the Fen-men for the drowning of the land? Did not the water bring them fish, and the fish attract wild fowl, which they could snare and shoot? Thus the proposal to drain the Fens and to convert them into wholesome and fruitful lands, how-ever important in a national point of view, as enlarging the resources and increasing the wealth of the country, had no attraction whatever in the eyes of the Fen-men. They mut-tered their discontent, and everywhere met the "adven-turers," as the reclaimers were called, with angry though ineffectual opposition. But their numbers were too few, and they were too widely scattered, to make any combined effort at resistance. They could only retreat to other fens where they thought they might still be safe, carrying their discon-tent with them, and complaining that their commons were taken from them by the rich, and, what was worse, by foreigners—Dutch and Flemings. The jealous John Bull of the towns became alarmed at this idea, and had rather that the water than these foreigners had possession of the land. " What !" asked one of the objectors, " is the old activitie and abilities of the English nation grown now soe dull and insufficient that wee must pray in ayde of our neighbours to improve our own demaynes? For matter of securitie, shall wee esteem it of small moment to put into the hands of strangers three or four such ports as Linne, Wisbeach, Spalding, and Boston, and permit the countrie within and between them to be peopled with overthwart neighbours ; or, if they quaile themselves, must wee give place to our most auncient and daungerous enemies, who will be readie enough to take advantage of soe manie fair inlets into the bosom of our land, lying soe near together that an army landing in each of them may easily meet and strongly entrench themselves with walls of water, and drowne the countrie about them at their pleasure ? " *

---

* ' The Drayner Confirmed,' tract, 1629.

Thus a great agitation against the drainage sprang up in the Fen districts, and a wide-spread discontent prevailed, which, as we shall afterwards find, exercised an important influence on the events which culminated in the Great Rebellion of a few years later. Among the other agencies brought to bear against the Fen drainers was the publication of satirical songs and ballads—the only popular press of the time; and the popular poets doubtless represented accurately enough the then state of public opinion, as their ballads were sung with great applause about the streets of the Fen towns. One of these, entitled 'The Powte's * Complaint,' was among the most popular.

* Powte—the old English word for the sea-lamprey.

THE POWTE'S COMPLAINT.

Come, Brethren of the water, and let us all assemble,
To treat upon this Matter, which makes us quake and tremble;
For we shall rue, if it be true that Fens be undertaken,
And where we feed in Fen and Reed, they'll feed both Beef and Bacon.

They'll sow both Beans and Oats, where never Man yet thought it;
Where Men did row in Boats, ere Undertakers bought it;
But, *Ceres*, thou behold us now, let wild Oats be their Venture,
Oh, let the Frogs and miry Bogs destroy where they do enter.

Behold the great Design, which they do now determine,
Will make our Bodies pine, a prey to Crows and Vermine;
For they do mean all Fens to drain, and Waters overmaster,
All will be dry, and we must die—'cause Essex Calves want pasture.

Away with Boats and Rudder, farewel both Boots and Skatches,
No need of one nor t'other, Men now make better Matches;
Stilt-makers all, and Tanners, shall complain of this Disaster,
For they will make each muddy lake for Essex Calves a Pasture.

The feather'd Fowls have Wings, to fly to other Nations;
But we have no such things, to help our Transportations;
We must give place, O grevious Case ! to horned Beasts and Cattle,
Except that we can all agree to drive them out by Battel.

Wherefore let us intreat our antient Water-Nurses
To shew their Power so great as t' help to drain their Purses;
And send us good old Captain Flood to lead us out to Battel,
Then Two-penny Jack, with Scales on 's Back, will drive out all the Cattle.

This Noble Captain yet was never know to fail us,
But did the conquest get of all that did assail us;
His furious Rage none could assuage; but, to the World's great Wonder,
He tears down Banks, and breaks their Cranks and Whirligigs assunder.

God *Eolus*, we do thee pray that thou wilt not be wanting;
Thou never said'st us nay—now listen to our canting;
Do thou deride their Hope and Pride that purpose our Confusion,
And send a Blast that they in haste may work no good Conclusion.

Great

In another popular drinking song, entitled 'The Drain-ing of the Fennes,' the Dutchmen are pointed out as the great offenders. The following stanzas may serve as a specimen :—

> " The Dutchman hath a thirsty soul,
>   Our cellars are subject to his call ;
> Let every man, then, lay hold on his bowl,
>   'Tis pity the German sea should have all.
> Then apace, apace drink, drink deep, drink deep,
>   Whilst 'tis to be had let's the liquor ply ;
> The drainers are up, and a coile they keep,
>   And threaten to drain the kingdom dry.
>
> Why should we stay here, and perish with thirst ?
>   To th' new world in the moon away let us goe,
> For if the Dutch colony get thither first,
>   'Tis a thousand to one but they'll drain that too!
>   *Chorus*—Then apace, apace drink, &c.

The Fen drainers might, however, have outlived these attacks, had the works executed by them been successful ; but unhappily they failed in many respects. Notwith-standing the numerous deep cuts made across the Fens in all directions at such great cost, the waters still retained their hold upon the land. The Bedford River and the other drains merely acted as so many additional receptacles for the surplus water, without relieving the drowned dis-tricts to any appreciable extent. This arose from the engineer confining his attention almost exclusively to the inland draining and embankments, while he neglected to provide any sufficient outfalls for the waters themselves into the sea. Vermuyden committed the error of adopting

---

Great *Neptune*, God of Seas, this Work must needs provoke ye ;
They mean thee to disease, and with Fen-Water choak thee ;
But with thy Mace do thou deface, and quite confound this matter,
And send thy Sands to make dry lands when they shall want fresh Water.

And eke we pray thee, *Moon*, that thou wilt be propitious,
To see that nought be done to prosper the Malicious ;
Tho' Summer's Heat hath wrought a Feat, whereby themselves they flatter
Yet be so good as send a Flood, lest Essex Calves want Water.

the Dutch method of drainage, in a district where the circumstances differed in many material respects from those which prevailed in Holland. In Zeeland, for instance, the few rivers passing through it were easily banked up and carried out to sea, whilst the low-lying lands were kept clear of surplus water by pumps driven by windmills. There, the main object of the engineer was to build back the river and the ocean; whereas in the Great Level the problem to be solved was, how to provide a ready outfall to the sea for the vast body of fresh water falling upon as well as flowing through the Fens themselves. This essential point was unhappily overlooked by the early drainers; and it has thus happened that the chief work of modern engineers has been to rectify the errors of Vermuyden and his followers; more especially by providing efficient outlets for the discharge of the Fen waters, deepening and straightening the rivers, and compressing the streams in their course through the Level, so as to produce a more powerful current and scour, down to their point of outfall into the sea.

This important condition of successful drainage having been overlooked, it may readily be understood how unsatisfactory was the result of the works first carried out in the Bedford Level. In some districts the lands were no doubt improved by the additional receptacles provided for the surplus waters, but the great extent of fen land still lay for the most part wet, waste, and unprofitable. Hence, in 1634, a Commission of Sewers held at Huntingdon pronounced the drainage to be defective, and the 400,000 acres of the Great Level to be still subject to inundation, especially in the winter season. The King, Charles I., then resolved himself to undertake the reclamation, with the object of converting the Level, if possible, into " winter grounds." He took so much personal interest in the work that he even designed a town to be called Charleville, which was to be built in the midst of the Level, for the purpose of commemorating the undertaking. Sir Cornelius

Vermuyden was again employed, and he proceeded to carry out the King's design. He had many enemies, but he could not be dispensed with; being the only man of recognised ability in works of drainage at that time in England.

The works constructed in pursuance of this new design were these:—an embankment on the south side of Morton's Leam, from Peterborough to Wisbeach; a navigable sasse, or sluice, at Standground; a new river cut between the stone sluice at the Horse-shoe and the sea below Wisbeach, 60 feet broad and 2 miles long, embanked at both sides; and a new sluice in the marshes below Tydd, upon the outfall of Shire Drain. These and other works were in full progress, when the political troubles of the time came to a height, and brought all operations to a stand-still for many years. The discontent caused throughout the Fens by the drainage operations had by no means abated; but, on the contrary, considerably increased. In other parts of the kingdom, the attempts made about the same time by Charles I. to levy taxes without the authority of Parliament gave rise to much agitation. In 1637 occurred Hampden's trial, arising out of his resistance to the payment of ship-money: by the end of the same year the King and the Parliamentary party were mustering their respective forces, and a collision between them seemed imminent.

At this juncture the discontent which prevailed throughout the Fen counties was an element of influence not to be neglected. It was adroitly represented that the King's sole object in draining the Fens was merely to fill his impoverished exchequer, and enable him to govern without a Parliament. The discontent became fanned into a fierce flame; on which Oliver Cromwell, the member for Huntingdon, until then comparatively unknown, availing himself of the opportunity which offered, of increasing the influence of the Parliamentary party in the Fen counties, immediately put himself at the head of a vigorous agitation against the

further prosecution of the scheme. He was very soon the
most popular man in the district; he was hailed 'Lord of
the Fens' by the Fen-men; and he went from meeting to
meeting, stirring up the public discontent, and giving it a
suitable direction. "From that instant," says Mr. Forster,*
"the scheme became thoroughly hopeless. With such
desperate determination he followed up his purpose—so
actively traversed the district, and inflamed the people
everywhere—so passionately described the greedy claims of
royalty, the gross exactions of the commission, nay, the
questionable character of the improvement itself, even could
it have gone on unaccompanied by incidents of tyranny,—
to the small proprietors insisting that their poor claims
would be merely scorned in the new distribution of the
property reclaimed,—to the labouring peasants that all the
profit and amusement they had derived from *commoning* in
those extensive wastes were about to be snatched for ever
from them,—that, before his almost individual energy,
King, commissioners, noblemen-projectors, all were forced to
retire, and the great project, even in the state it then was,
fell to the ground."

The success of the Cambridge Fen-men, in resisting the
reclamation of the wastes, encouraged those in the more
northern districts to take even more summary measures to
get rid of the drainers, and restore the lands to their former
state. The Earl of Lindsey had succeeded at great cost in
enclosing and draining about 35,000 acres of the Lindsey
Level, and induced numerous farmers and labourers to
settle upon the land. They erected dwellings and farm-
buildings, and were busily at work, when the Fen-men
suddenly broke in upon them, destroyed their buildings,
killed their cattle, and let in the waters again upon the
land. So, too, in the West and Wildmore Fen district
between Tattershall and Boston in Lincolnshire, where

---

* Lives of Eminent British Statesmen (Lardner's 'Cabinet Cyclopædia,'
vol. vi. p. 60).

considerable progress had been made by a body of "adventurers" in reclaiming the wastes. After many years' labour and much cost, they had succeeded in draining, enclosing, and cultivating an extensive tract of rich land, and they were peaceably occupied with their farming pursuits, when a mob of Fen-men collected from the surrounding districts, and under pretence of playing at football, levelled the enclosures, burnt the corn and the houses, destroyed the cattle, and even killed many of the people who occupied the land. They then proceeded to destroy the drainage works, by cutting across the embankments and damming up the drains, by which the country was again inundated and restored to its original state.

The greater part of the Level thus again lay waste, and the waters were everywhere extending their dominion over the dry land through the choking up of the drains and river outfalls by the deposit of silt. Matters were becoming even worse than before, but could not be allowed thus to continue. In 1641 the Earl of Bedford and his participants made an application to the Long Parliament, then sitting, for permission to re-enter upon the works; but the civil commotions which still continued prevented any steps being taken, and the Earl himself shortly after died in a state of comparative penury, to which he had reduced himself by his devotion to this great work. Again, however, we find Sir Cornelius Vermuyden upon the scene. Undaunted by adversity, and undismayed by the popular outrages committed upon his poor countrymen in Lincolnshire and Yorkshire, he still urged that the common weal of England demanded that the rich lands lying under the waters of the Fens should be reclaimed, and made profitable for human uses. He saw a district almost as large as the whole of the Dutch United Provinces remaining waste and worse than useless, and he gave himself no rest until he had set on foot some efficient measure for its drainage and reclamation. What part he took in the political discussions of the time, we know not; but we find the eldest of his sons, Cornelius, a colonel in the Parliamentary army * stationed in the Fens under Fairfax, shortly before the battle of Naseby. Vermuyden himself was probably too much engrossed by his drainage project to give heed to political affairs; and besides, he could not forget that Charles, and Charles's father, had been his fast friends.

In 1642, while the civil war was still raging, appeared Vermuyden's 'Discourse' on the Drainage of the Fens, wherein he pointed out the works which still remained to be executed in order effectually to reclaim the 400,000 acres of land capable of growing corn, which formed the area of the Great Level. His suggestions formed the subject of much pamphleteering discussion for several years, during which also numerous petitions were presented to Parliament, urging the necessity for perfecting the drainage. At length, in 1649, authority was granted to William, Earl of Bedford, and other participants, to prosecute the undertaking which his father had begun, and steps were shortly after taken to recommence the works. Again was Westerdyke, the Dutch engineer, called in to criticise Vermuyden's plans; and again was Vermuyden triumphant over his opponent. He was selected, once more, to direct the drainage, which, looking at the defects of the works previously executed by him, and the difficulties in which the first Earl had thereby become involved, must be regarded as a marked proof of the man's force of purpose, as well as of his recognised integrity of character.

Vermuyden again collected his Dutchmen about him, and vigorously began operations. But they had not proceeded far before they were again almost at a standstill for want

---

* "The party under Vermuyden waits the King's army, and is about Deeping; has a command to join with Sir John Gell, if he commands him."—Cromwell's Letter to Fairfax, 4th June, 1645. This Vermuyden resigned his commission a few days before the battle of Naseby, having, as he alleged, special reasons requiring his presence beyond the seas, whence he does not seem to have returned until after the Restoration. In 1665 we find him a member of the Corporation of the Bedford Level.

of funds; and throughout their entire progress they were hampered and hindered by the same great difficulty. Some of the participants sold and alienated their shares in order to get rid of further liabilities ; others held on, but became reduced to the lowest ebb. Means were, however, adopted to obtain a supply of cheaper labour; and application was made by the adventurers for a supply of men from amongst the Scotch prisoners who had been taken at the battle of Dunbar. A thousand of them were granted for the purpose, and employed on the works to the north of Bedford River, where they continued to labour until the political arrangements between the two countries enabled them to return home. When the Scotch labourers had left, some difficulty was again experienced in carrying on the works. The local population were still hostile, and occasionally interrupted the labourers employed upon them; a serious riot at Swaffham having only been put down by the help of the military. Blake's victory over Van Tromp, in 1652, opportunely supplied the Government with a large number of Dutch prisoners, five hundred of whom were at once forwarded to the Level, where they proved of essential service as labourers.

The most important of the new rivers, drains, and sluices included in this further undertaking, were the following :—
The New Bedford River, cut from Erith on the Ouse to Salter's Lode on the same river, reducing its course between these points from 40 to 20 miles : this new river was 100 feet broad, and ran nearly parallel with the Old Bedford River. A high bank was raised along the south side of the new cut, and an equally high bank along the north side of the old river, a large space of land, of about 5000 acres, being left between them, called the Washes, for the floods to "bed in," as Vermuyden termed it. Then the river Welland was defended by a bank, 70 feet broad and 8 feet high, extending from Peakirk to the Holland bank. The river Nene was also defended by a similar bank, extending from Peterborough to Guyhirne; and another bank was raised

between Standground and Guyhirne, so as to defend the
Middle Level from the overflowing of the Northamptonshire
waters. The river Ouse was in like manner restrained by high
banks extending from Over to Erith, where a navigable
sluice was provided. Smith's Leam was cut, by which the
navigation from Wisbeach to Peterborough was opened out.
Among the other cuts and drains completed at the same time,
were Vermuyden's Eau, or the Forty Feet Drain, extending
from Welch's Dam to the river Nene near Ramsey Mere;
Hammond's Eau, near Somersham, in the county of Huntingdon; Stonea Drain and Moore's Drain, near March, in
the Isle of Ely; Thurlow's Drain, extending from the
Forty Feet to Popham's Eau; and Conquest Lode, leading
to Whittlesea Mere. And in order to turn the tidal waters
into the Hundred Feet River, as well as to prevent the
upland floods from passing up the Ten Mile River towards
Littleport, Denver Sluice, that great bone of after conten-
tion, was constructed. Another important work in the South
Level was the cutting of a large river called St. John's, or
Downham Eau,* 120 feet wide, and 10 feet deep, from
Denver Sluice to Stow Bridge on the Ouse, with sluices at
both ends, for the purpose of carrying away with greater
facility the flood waters descending from the several rivers
of that level. Various new sluices were also fixed at the
mouths of the rivers, to prevent the influx of the tides, and
most of the old drains and cuts were at the same time
scoured out and opened for the more ready flow of the
surface waters.

At length, in March, 1652, the works were declared to
be complete, and the Lords Commissioners of Adjudication
appointed under the Act of Parliament proceeded to inspect
them. They embarked upon the New River, and sailing
over it to Stow Bridge, surveyed the new eaus and sluices
executed near that place, after which they returned to Ely

* The St. John's Eau, being a
straight cut, is known in the district
as "The Poker;" and Marshland
Cut, being in the shape of a pair o
tongs, is commonly called "Tong
Drain."

There Sir Cornelius Vermuyden read to those assembled a discourse, in which he explained the design he had carried out for the drainage of the district; in the course of which he stated as one of the results of the undertaking, that in the North and Middle Levels there were already 40,000 acres of land " sown with cole-seed, wheat, and other winter grain, besides innumerable quantities of sheep, cattle, and other stock, where never had been any before. These works," he added, " have proved themselves sufficient, as well by the great tide about a month since, which overflowed Marshland banks, and drowned much ground in Lincoln-shire and other places, and a flood by reason of a great snow, and rain upon it following soon after, and yet never hurt any part of the whole Level; and the view of them, and the consideration of what hath previously been said, proves a clear draining according to the Act." He con-cluded thus,—" I presume to say no more of the work, lest I should be accounted vain-glorious; although I might truly affirm that the present or former age have done nothing like it for the general good of the nation. I humbly desire that God may have the glory, for his blessing and bringing to perfection my poor endeavours, at the vast charge of the Earl of Bedford and his participants."

A public thanksgiving took place to celebrate the com-pletion of the undertaking; and on the 27th of March, 1653, the Lords Commissioners of Adjudication of the Reclaimed Lands, accompanied by their officers and suite,—the Com-pany of Adventurers, headed by the Earl of Bedford,—the magistrates and leading men of the district, with a vast concourse of other persons,—attended public worship in the cathedral of Ely, when the Rev. Hugh Peters, chaplain to the Lord-General Cromwell, preached a sermon on the occasion.

Vermuyden's perseverance had thus far triumphed. He had stood by his scheme when all others held aloof from it. Amidst the engrossing excitement of the civil war, the one dominating idea which possessed him was the drainage of the Great Level. While the nation was divided into two hostile camps, and the deadly struggle was proceeding between the Royalists and the Parliamentarians, Vermuyden's sole concern was how to raise the funds wherewith to pay his peaceful army of Dutch labourers in the Fens. To carry on the works he sold every acre of the soil he had reclaimed. He first sold the allotment of land won by him from the Thames at Dagenham in 1621; then he sold his interest in his lands at Sedgemoor and Malvern Chase; and in 1654 we find him conveying the remainder of his property in Hatfield Level. He was also under the necessity of selling all the lands apportioned to him in the Bedford Level itself, in order to pay the debts incurred in their drainage. But although he lost all, it appears that the company in the end preferred heavy pecuniary claims against him which he had no means of meeting; and in 1656 we find him appearing before Parliament as a suppliant for redress. Thenceforward he entirely disappears from public sight; and it is supposed that, very shortly after, he went abroad and died, a poor, broken down old man, the extensive lands which he had reclaimed and owned having been conveyed to strangers.

The drainage of the Fens, however, was not yet complete. The district was no longer a boggy wilderness, but much of it in fine seasons was covered with waving crops of corn. As the swamps were drained, farm buildings, villages, and towns gradually sprang up, and the toil of the labourer was repaid by abundant harvests. The anticipation held forth in the original charter granted by Charles I. to the reclaimers of the Bedford Level was more than fulfilled. "In those places which lately presented nothing to the eyes of the beholders but great waters and a few reeds thinly scattered here and there, under the Divine mercy might be seen pleasant pastures of cattle and kine, and many houses belonging to the inhabitants." But the tenure by which the land continued to be held was unremitting vigilance * and industry; the difficulties interposed by nature tending to discipline the skill, to stimulate the enterprise, and evoke the energy of the people who had rescued the fields from the watery waste.

Improvements of all kinds went steadily on, until all the rivers flowing through the Level were artificially banked and diverted into new channels, excepting the Nene, which is the only natural river in the Fen district remaining comparatively unaltered. New dykes, causeways, embankments, and sluices were formed; many droves, leams, eaus, and drains were cut, furnished with gowts or gates at their lower ends, which were from time to time dug, deepened, and widened. Mills were set to work to pump out the water from the low grounds; first windmills, sometimes with double-lifts, as practised in Holland; and more recently powerful steam-engines. Sluices were also erected to prevent the inland waters from returning; strong embankments extending in all directions, to keep the rivers and tides within their defined channels. To protect the land from the sea waters as well as the fresh,—to build and lock back the former, and to keep the latter within due limits,—was the work of the engineer; and by his skill, aided by the industry of his contractors and workmen, water, instead of being the master and tyrant as of old, became man's servant and pliant agent, and was used as an irrigator, a conduit, a mill-stream, or a water-road for ex-

---

* Since the above passage was written, in 1861, its truth has been amply confirmed by the blowing up of the Middle Level sluice, about three miles above Lynn, by which some 10,000 acres of the richest agricultural land in Marshland were completely submerged. Great loss was thereby occasioned, and a fertile crop of lawsuits has followed the inundation. The Middle Level Drain was admirably planned by the late James Walker, C.E., and was sup- posed to have been as admirably executed, at a cost of over 400,000*l*. For ten years it was a complete success; but the foundations of the outfall sluice having become undermined by the tidal scour in the Ouse, the masonry fell in, and the waters immediately rushed in upon the land and resumed possession. The drainage has since been restored in a very able and efficient manner by Mr. Hawkshaw, and Marshland is again under cultivation.

tensive districts of country. In short, in no part of the world, except in Holland, have more industry and skill been displayed in reclaiming and preserving the soil, than in Lincolnshire and the districts of the Great Bedford Level. Six hundred and eighty thousand acres of the most fertile land in England, or an area equal to that of North and South Holland, have been converted from a dreary waste into a fruitful plain, and fleets of vessels traverse the district itself, freighted with its rich produce. Taking its average annual value at 4*l.* an acre, the addition to the national wealth and resources may be readily calculated.

The prophecies of the decay that would fall upon the country, if " the valuable race of Fenmen " were deprived of their pools for pike, and fish, and wild-fowl, have long since been exploded. The population has grown in numbers, in health, and in comfort, with the progress of drainage and reclamation. The Fens are no longer the lurking places of disease,* but as salubrious as any other parts of England. Dreary swamps are supplanted by pleasant pastures, and the haunts of pike and wild-fowl have become the habitations of industrious farmers and husbandmen. Even Whittlesea Mere and Ramsey Mere, —the only two lakes, as we were told in the geography books of our younger days, to be found in the south of England,—have been blotted out of the map, for they have been drained by the engineer, and are now covered with smiling farms and pleasant homesteads.

---

* When Dr. Whalley was presented by the Bishop of Ely to the rectory of Hagworthingham-in-the-Fens, it was with the singular proviso that he was not to reside in it, as the air was fatal to any but a native. (*Journals and Correspondence of T. S. Whalley, D.D.*) Statistics, however, prove that the Fen districts are the exceptional haunts of disease. The Registrar-General, in one of his recent reports, states that, whilst the mortality of Pau in the Pyrenees, a place resorted to by British invalids on account of its salubriousness, is 23 in 1000, that of Ely is only 17 in 1000.

CHAPTER III.

SIR HUGH MYDDELTON. — THE CUTTING OF THE NEW RIVER.

WHILE the engineer has occasionally to contend with all his skill against the powers of water, he has also to deal with it as a useful agent. Though water, like fire, is a bad master, the engineer contrives to render it docile and tractable. He leads it in artificial channels for the purpose of driving mills and machinery, or he employs it to feed canals along which boats and ships laden with merchandise may be safely floated.

But water is also an indispensable necessary of life, an abundant supply of it being essential for human health and comfort. Hence nearly all the ancient towns and cities were planted by the banks of rivers, principally because the inhabitants required a plentiful supply of water for their daily uses. Old London had not only the advantage of its pure broad stream flowing along its southern boundary, so useful as a water-road, but it also possessed an abundance of Wells, from which a supply of pure water was obtained, adequate for the requirements of its early population. The river of Wells, or Wallbrook, flowed through the middle of the city; and there were numerous wells in other quarters, the chief of which were Clerke's Well, Clement's Well, and Holy Well, the names of which still survive in the streets built over them.

As London grew in size and population, these wells were found altogether inadequate for the wants of the inhabitants; besides, the water drawn from them became tainted by the impurities which filter into the soil wherever large numbers are congregated. Conduits were then constructed, through which water was led from Paddington, from James's Head, Mewsgate, Tyburn, Highbury, and Hampstead. There were sixteen of such public conduits about London, and the Conduit Streets which still exist through-

Ancient Conduit in Westcheap.

out the metropolis mark the sites of several of these ancient works.* The copious supply of water by the conduits was

---

* The conduits used, in former times, to be yearly visited with considerable ceremony. For instance, we find that—" On the 18th of September, 1562, the Lord Mayor (Harper), the Aldermen, with many worshipful persons, and divers of the Masters and Wardens of the twelve companies, rode to the Conduit's-head [now the site of Conduit Street, New Bond Street], for to see them after the old custom. And afore dinner they hunted the hare and killed her, and thence to dinner at the head of the Conduit. There was a good number entertained with good cheere by the Chamberlain, and, after dinner, they hunted the fox. There was a great cry for a mile, and at length the hounds killed him at the end of St. Giles's. Great hallooing at his death, and blowing of hornes; and thence the Lord Mayor, with all his company, rode through London to his place in Lombard Street." Stow's 'Survey of London.'—It would appear that the ladies of the Lord Mayor and Aldermen attended on these jovial occasions, riding *in waggons*.

all the more necessary at that time, as London was for
the most part built of timber, and liable to frequent fires,
to extinguish which promptly, every citizen was bound to
have a barrel full of water in readiness outside his door.
The corporation watched very carefully over their protection, and inflicted severe punishments on such as interfered
with the flow of water through them. We find a curious
instance of this in the City Records, from which it appears
that, on the 12th November, 1478, one William Campion,
resident in Fleet Street, had cunningly tapped the conduit
where it passed his door, and conveyed the water into a
well in his own house, " thereby occasioning a lack of water
to the inhabitants." Campion was immediately had up
before the Lord Mayor and Aldermen, and after being confined for a time in the Comptour in Bread Street, the following further punishment was inflicted on him. He was
set upon a horse with a vessel like unto a conduit placed
upon his head, which being filled with water running out
of small pipes from the same vessel, he was taken round all
the conduits of the city, and the Lord Mayor's proclamation
of his offence and the reason for his punishment was then
read. When the conduit had run itself empty over the
culprit, it was filled again. The places at which the proclamation was read were the following,—at Leadenhall, at
the pillory in Cornhill, at the great conduit in Chepe, at
the little conduit in the same street, at Ludgate and Fleet
Bridge, at the Standard in Fleet Street, at Temple Bar, and
at St. Dunstan's Church in Fleet Street; from whence he
was finally marched back to the Comptour, there to abide
the will of the Lord Mayor and Aldermen.*

But the springs from which the conduits were supplied in
course of time decayed; perhaps they gradually diminished
by reason of the sinking of wells in their neighbourhood
for the supply of the increasing suburban population.
Hence a deficiency of water began to be experienced in the

---

* 'Corporation Records,' Index No. I., fo. 184 b.

city, which in certain seasons almost amounted to a famine.
There were frequent contentions at the conduits for "first
turn," and when water was scarce, these sometimes grew
into riots. The water carriers came prepared for a fight,
and at length the Lord Mayor had to interfere, and issued
his proclamation forbidding persons from resorting to the
conduits armed with clubs and staves. This, however, did
not remedy the deficiency. It is true the Thames,—"that
most delicate and serviceable river," as Nichols terms it,*
was always available; but an increasing proportion of
the inhabitants lived at a distance from the river. Besides,
the attempt was made by those who occupied the lanes
leading towards the Thames to stop the thoroughfare, and
allow none to pass without paying a toll. A large number
of persons then obtained a living as water carriers,† selling
the water by the "tankard" of about three gallons; and
they seem to have formed a rather unruly portion of the
population.

The difficulty of supplying a sufficient quantity of water
to the inhabitants by means of wells, conduits, and water
carriers, continued to increase, until the year 1582, when
Peter Morice, the Dutchman, undertook, as the inhabitants
could not go to the Thames for their water, to carry the
Thames to them. With this object he erected an ingenious
pumping engine in the first arch of London Bridge, worked

---

* 'Progresses of James I.,' vol. ii.,
699. The Corporation records contain numerous references to the preservation of the purity of the water
in the river. The Thames also furnished a large portion of the food of
the city, and then abounded in salmon and other fish, the London
fishermen constituting a large class.
We find numerous proclamations
issued relative to the netting of the
"salmon and porpoises," wide nets
and wall nets being especially pro-
hibited. Fleets of swans on the
Thames were a picturesque feature

of the river down even to the time
of James II.
    † The water carrier was com-
monly called a "Cob," and Ben
Jonson seems to have given a sort
of celebrity to the character by his
delineation of "Cob" in his 'Every
Man in his Humour.' Gifford, in
a note on the play, pointed out that
there is an avenue still called
"Cob's Court," in Broadway, Black-
friars; not improbably (he adds)
from its having formerly been inhabited principally by the class of
water carriers.

E 2

by water wheels driven by the rise and fall of the tide, which then rushed with great velocity through the arches. This machine forced the water through leaden pipes, laid into the houses of the citizens. The power with which Morice's forcing pumps worked was such, that he was enabled to throw the water over St. Magnus's steeple, greatly to the astonishment of the Mayor and Aldermen, who assembled to witness the experiment. The machinery succeeded so well that a few years later we find the corporation empowering the same engineer to use the second arch of London Bridge for a similar purpose.*

But even this augmented machinery for pumping was found inadequate for the supply of London. The town was extending rapidly in all directions, and the growing density of the population along the river banks was every year adding to the impurity of the water, and rendering it less and less fit for domestic purposes. Hence the demand for a more copious and ready supply of pure water continued steadily to increase. Where was the new supply to be obtained, and how was it to be rendered most readily available for the uses of the citizens? Water is by no means a scarce element in England; and no difficulty was experienced in finding a sufficiency of springs and rivers of pure water at no great distance from the metropolis. Thus, various springs were known to exist in different parts of Hertfordshire and Middlesex; and many vague projects were proposed for conveying their waters to London.

Desiring that one plan or another might be carried out, the corporation obtained an Act towards the end of Queen Elizabeth's reign,† empowering them to cut a river to the city from any part of Middlesex or Hertfordshire; and ten years were specified as the time for carrying out the necessary works. But, though many plans were sug-

---

* The river pumping-leases continued in the family of the Morices until 1701, when the then owner sold his rights to Richard Soams for 38,000l., and by him they were afterwards transferred to the New River Company at a still higher price.   † Act 13 Eliz. c. 18.

gested and discussed, no steps were taken to cut the proposed river. The enterprise seemed too large for any private individual to undertake; and though the corporation were willing to sanction it, they were not disposed to find any part of the requisite means for carrying it out. Notwithstanding, therefore, the necessity for a large supply of water, which became more urgent in proportion to the increase of population, the powers of the Act were allowed to expire without anything having been done to carry them into effect.

In order, however, to keep alive the parliamentary powers, another Act was obtained in the third year of James I.'s reign (1605),* to bring an artificial stream of pure water from the springs of Chadwell and Amwell, in Hertfordshire; and the provisions of this Act were enlarged and amended in the following session.† From an entry in the journals of the corporation, dated the 14th October, 1606, it appears that one William Inglebert petitioned the court for liberty to bring the water from the above springs to the northern parts of the city "in a trench or trenches of brick." The petition was "referred," but nothing further came of it; and the inhabitants of London continued for some time longer to suffer from the famine of water—the citizens patiently waiting for the corporation to move, and the corporation as patiently waiting for the citizens.

The same inconveniences of defective water-supply were experienced in other towns, and measures were in some cases taken to remedy them. Thus, at Hull, in certain seasons, the inhabitants were under the necessity of bringing the water required by them for ordinary uses across the Humber from Lincolnshire in boats, at great labour and expense. They sought to obtain a better supply by leading water into the town from the streams in the neighbourhood; but the villagers of Hessle, Anlaby, and Cottingham, with others, resisted their attempts. In 1376, the mayor and

---

* 3 Jac. c. 18 | † 4 Jac. c. 12.

burgesses appealed to the Crown; commissioners were appointed to inquire into the subject; and the result was, that powers were granted for making an aqueduct from Anlaby Springs to Hull. This was not accomplished without serious opposition on the part of the villagers, who riotously assembled to destroy the works, and they even went so far as to threaten Hull itself with destruction. Some of the rioters were seized and hanged at York, and the aqueduct was then finished.

Tiverton, in Devonshire, has in like manner been supplied with water from a very early period, by means of an artificial cut called the Town-leet, extending from a spring on White Down, about five miles distant, into the heart of the town. This valuable conduit was the free gift of Amicia, Countess of Devon, to the inhabitants, as long ago as the year 1240; and it continues a constant source of blessing. A perambulation is made along its course once in every five or six years, by the portreeve, the steward of the manor, the water bailiffs, and others, from Cogan's Well in the centre of the town, to the source of the stream on White Down. All obstructions are then removed, and the stream is claimed publicly for the sole use of the inhabitants of Tiverton. For about two miles of its course, it may, perhaps, be regarded as a natural stream; but from the village of Chettiscombe the channel is for the most part artificial, the water being confined within a high embankment, in many places above the level of the surrounding country; and it is conveyed, as one writer says, " over a deep road behind the hospital, by a leaded shute, on a strong stone arch, into the town." * By this channel, the water, shedding in its passage an allotted portion to each street, is brought to Cogan's Well, where it is artificially parted into three streams, which run along the sides of the remaining streets, until they are discharged into one or other of the two rivers, the Loman and the Exe, from which the place

---

* Dunsford's 'Historical Memoirs,' p. 106.

derives its name of Twy-ferd-town, or Tiverton. Copious rivulets are in like manner led, by artificial cuts, through the principal streets of Salisbury, from the natural streams at the confluence of which that city is situated.

But the most important artificial work of the kind in the West of England is that constructed for the supply of water to Plymouth, which was carried out through the public spirit and enterprise of one of the most distinguished of English admirals—the great Sir Francis Drake. It appears from ancient records that water was exceedingly scarce in that town, the inhabitants being under the necessity of sending their clothes more than a mile to be washed, the water used by them for domestic purposes having to be fetched for the most part from Plympton, about five miles distant. Sir Francis Drake, who was born within ten miles of Plymouth, and had settled in the neighbourhood, after having realized a considerable fortune by his adventures on the Spanish main, observing the great inconvenience suffered by the population from their want of water, as well as the difficulty of furnishing a supply to the ships frequenting the port, conceived the project of remedying the defect by leading a store of water to the town from one of the numerous springs on Dartmoor. Accordingly, in 1587, when he represented Bossiney (Tintagel) in Cornwall, he obtained an Act enabling him to convey a stream from the river Mew or Meavy; and in the preamble to the Act it was expressed that its object was not only to ensure a continual supply of water to the inhabitants, but to obviate the inconvenience hitherto sustained by seamen in watering their vessels. It would appear, from documents still extant, that the town of Plymouth contributed 200l. towards the expenses of the works, Sir Francis being at the remainder of the cost; and on the completion of the undertaking the corporation agreed to grant him a lease of the aqueduct for a term of twenty years, at a nominal rental. Drake lost no time in carrying out the work, which was finished in four years after the passing of the

Act; and its completion in 1591, on the occasion of the welcoming of the stream into the town, was celebrated by great public rejoicings.*

The "Leet," as it is called, is a work of no great magnitude, though of much utility. It was originally nothing more than an open trench cut along the sides of the moor, in which the water flowed by a gentle inclination into the town and through the streets of Plymouth. The distance between the head of the aqueduct at Sheep's Tor and Plymouth, as the crow flies, is only seven miles; but the length of the Leet—so circuitous are its windings—is nearly twenty-four miles. After its completion, Drake presented the aqueduct to the inhabitants of Plymouth "as a free gift for ever," and it has since remained vested in the corporation,—who might, however, bestow more care than they do on its preservation against impurity. Two years after the completion of the Leet, the burgesses, probably as a mark of their gratitude, elected Drake their representative in Parliament. The water proved of immense public convenience, and Plymouth, instead of being one of the worst supplied, was rendered one of the best watered towns in the kingdom. Until a comparatively recent date the water flowed from various public conduits, and it ran freely on either side of the streets, that all classes of the people might enjoy the benefit of a full and permanent supply throughout the year. One of the original conduits still remains at the head of Old Town-street, bearing the inscription, "Sir Francis Drake first brought this water into Plymouth, 1591."

The example of Plymouth may possibly have had an influence upon the corporation of London in obtaining the

---

* The tradition survives to this day that Sir Francis Drake did not cut the Leet by the power of money and engineering skill, but by the power of magic. It is said of him that, calling for his horse, he mounted it and rode about Dartmoor until he came to a spring sufficiently copious for his design, on which, pronouncing some magical words, he wheeled round, and, starting off at a gallop, the stream formed its own channel, and followed his horse's heels into the town.

requisite powers from Parliament to enable them to bring the springs of Chadwell and Amwell to the thirsty population of the metropolis; but unhappily they had as yet no Drake to supply the requisite capital and energy.   In March, 1608, one Captain Edmond Colthurst petitioned the Court of Aldermen for permission to enter upon the work; but it turned out that the probable cost was far beyond the petitioner's means, without the pecuniary help of the corporation; and that being withheld, the project fell to the ground.   After this, one Edward Wright is said to have actually begun the works; but they were suddenly suspended, and fever and plague* continued to decimate the population.   The citizens of London seemed to be as far as ever from their supply of pure water.   At this juncture, when all help seemed to fail, and when men were asking each other "who is to do this great work, and how is it to be done?" citizen Hugh Myddelton, impatient of further delay, came forward and boldly offered to execute it at his own cost.   Yet Hugh Myddelton was not an engineer, nor even an architect nor a builder.   What he really was, we now proceed briefly to relate, according to the best information that we have been able to bring together on the subject.

Hugh Myddelton, the London goldsmith, was born in the year 1555, at Galch-hill, near Denbigh, in North Wales.

---

* The plague was then a frequent visitor in the city.   Numerous proclamations were made by the Lord Mayor and Corporation on the subject,—proclamations ordering wells and pumps to be drawn, and streets to be cleaned,—and precepts for removing hogs out of London, and against the selling or eating of pork. Wherever the plague was in a house, the inhabitant thereof was enjoined to set up outside a pole of the length of seven feet, with a bundle of straw at the top, as a sign that the deadly visitant was within.   Wife, children, and servants belonging to that house must carry white rods in their hands for thirty-six days before they were considered "purged."   It was also ordered subsequently, that on the street-door of every house infected, or upon a post thereby, the inhabitant must exhibit imprinted on paper a token of St. Anthony's Cross, otherwise called the sign of the Taw **T**, that all persons might have knowledge that such house was infected.—'Corporation of City of London Records,' jor. 12, fol. 136. No. 1.   Years 1590 to 1694.

Richard Myddelton, of Galch-hill, was governor of Denbigh Castle in the reigns of Edward VI., Mary, and

Myddelton's House at Galch-hill, Denbigh.

Elizabeth. He was a man eminent for his uprightness and integrity, and is supposed to have been the first member who sat in Parliament for the town of Denbigh. His wife was one Jane Dryhurst, the daughter of an alderman of the town, by whom he had a family of nine sons and seven daughters. He was buried with his wife in the parish church of Denbigh, called Whitchurch or St. Marcellus; where a small monumental brass, placed within the porch, represents Richard Myddelton and Jane his wife, with their sixteen children behind them, all kneeling.

Several of the Governor's sons rose to distinction. The third son, William, was one of Queen Elizabeth's famous

sea captains. He was also a man of literary tastes, being
the author of a volume entitled 'Barddonnaeth, or the Art of

Fac-simile of the Myddelton Brass in Whitchurch Porch.

Welsh Poetry.' While on his cruises, he occupied himself
in translating the Book of Psalms into Welsh; he finished
it in the West Indies, and it was published in 1603, shortly
after his death. The fourth son, Thomas, was an eminent
citizen and grocer of London. He served the office of
Sheriff in 1603, when he was knighted; and he was elected
Lord Mayor in 1613. He was the founder of the Chirk
Castle family, now represented by Mr. Myddelton Biddulph.
The fifth son, Charles, succeeded his father as governor of
Denbigh Castle, and when he died bequeathed numerous
legacies for charitable uses. The sixth son was Hugh, the
subject of this memoir. Robert, the seventh, was, like two
of his brothers, a citizen of London, and afterwards a mem-
ber of Parliament. Foulk, the eighth son, served as high

sheriff of the county of Denbigh. This was certainly a large measure of worldly prosperity and fame to fall to the lot of one man's offspring.

Hugh Myddelton was sent up to London to be bred to business there, under the eye of his elder brother Thomas, the grocer and merchant adventurer. In those days country gentlemen of moderate income were accustomed to bind their sons apprentices to merchants, especially where the number of younger sons was large, as it certainly was in the case of Richard Myddelton of Galch-hill. There existed at that time in the metropolis numerous exclusive companies or guilds, the admission into which was regarded as a safe road to fortune. The merchants were few in number, constituting almost an aristocracy in themselves; indeed, they were not unfrequently elevated to the peerage because of their wealth as well as public services, and not a few of our present noble families can trace their pedigree back to some wealthy skinner, mercer, or tailor, of the reigns of James or Elizabeth.

Hugh Myddelton was entered an apprentice of the guild of the Goldsmiths' Company. Having thus set his son in the way of well-doing, Richard Myddelton left him to carve out his own career, relying upon his own energy and ability. He had done the same with Thomas, whom he had helped until he could stand by himself; and William, whom he had educated at Oxford as thoroughly as his means would afford. These sons having been fairly launched upon the world, he bequeathed the residue of his property to his other sons and daughters.

The goldsmiths of that day were not merely dealers in plate, but in money. They had succeeded to much of the business formerly carried on by the Jews and Venetian merchants established in or near Lombard-street. They usually united to the trade of goldsmith that of banker, money-changer, and money-lender, dealing generally in the precious metals, and exchanging plate and foreign coin for gold and silver pieces of English manufacture, which had become much depreciated by long use as well as by frequent debasement. It was to the goldsmiths that persons in want of money then resorted, as they would now resort to money-lenders and bankers; and their notes or warrants of deposit circulated as money, and suggested the establishment of a bank-note issue, similar to our present system of bullion and paper currency. They held the largest proportion of the precious metals in their possession; hence, when Sir Thomas Gresham, one of the earliest bankers, died, it was found that the principal part of his wealth was comprised in gold chains.*

The place in which Myddelton's goldsmith's shop was situated was in Bassishaw (now called Basinghall) Street, and he lived in the overhanging tenement above it, as was then the custom of city merchants. Few, if any, lived away from their places of business. The roads into the country, close at hand, were impassable in bad weather, and dangerous at all times. Basing Hall was only about a bow-shot from the City Wall, beyond which lay Finsbury Fields, the archery ground of London, which extended from the open country to the very wall itself, where stood Moor Gate. The London of that day consisted almost exclusively of what is now called The City; and there were few or no buildings east of Aldgate, north of Cripplegate, or west of Smithfield. At the accession of James I. there were only a few rows of thatched cottages in the Strand, along which, on the river's side, the boats lay upon the beach. At the same time there were groves of trees in Finsbury and green pastures in Holborn; Clerkenwell was a village; St. Pancras boasted only of a little church standing in meadows; and St. Martin's, like St. Giles's, was literally "in the fields." All the country to the west was farm and pasture land; and woodcocks and partridges flew over the site of the future Regent Street, May Fair, and Belgravia.

---

* It may be remembered that Rubens was accustomed to be paid for his pictures by so many links of gold chain.

The population of the city was about 150,000, living in some 17,000 houses, brick below and timber above, with picturesque gable-ends, and sign boards swinging over the footways. The upper parts of the houses so overhung the foundations, and the streets were so narrow, that D'Avenant said the opposite neighbours might shake hands without stirring from home. The ways were then quite impassable for carriages, which had not yet indeed been introduced into England; all travelling being on foot or on horseback. When coaches were at length introduced and became fashionable, the aristocracy left the city, through the streets of which their carriages could not pass, and migrated westward to Covent Garden and Westminster.

Those were the days for quiet city gossip and neighbourly chat over matters of local concern; for London had not yet grown so big or so noisy as to extinguish that personal interchange of views on public affairs which continues to characterise most provincial towns. Merchants sat at their doorways in the cool of the summer evenings, under the overhanging gables, and talked over the affairs of trade; whilst those courtiers who still had their residences within the walls, lounged about the fashionable shops to hear the city gossip and talk over the latest news. Myddelton's shop appears to have been one of such fashionable places of resort, and the pleasant tradition was long handed down in the parish of St. Matthew, Friday-street, that Hugh Myddelton and Walter Raleigh used to sit together at the door of the goldsmith's shop, and smoke the newly introduced weed, tobacco, greatly to the amazement of the passers by. It is not improbable that Captain William Myddelton, who lived in London * after his return from the Spanish main in 1591, formed an occasional member of the group; for Pennant states that he and his friend Captain Thomas Price, of Plâsgollen, and another, Captain Koet, were the first who

---

* He resided at the old Elizabethan house in Highgate, afterwards occupied as an inn, called the "King's Head."

smoked, or as they then called it, "drank" tobacco publicly in London, and that the Londoners flocked from all parts to see them.[*]

Hugh Myddelton did not confine himself to the trade of a goldsmith, but from an early period his enterprising spirit led him to embark in ventures of trade by sea; and hence, when we find his name first mentioned in the year 1597, in the records of his native town of Denbigh, of which he was an alderman and "capitall burgess," as well as the representative in Parliament, he is described as "Cittizen and Gouldsmythe of London, and one of the Merchant Adventurers of England."[†] The trade of London was as yet very small, but a beginning had been made. A charter was granted by Henry VII., in 1505, to the Company of Merchant Adventurers of England, conferring on them special privileges. Previous to that time, almost the whole trade had been monopolised by the Steelyard Company of Foreign Merchants, whose exclusive privileges were formally withdrawn in 1552. But for want of an English mercantile navy, the greater part of the foreign carrying trade of the country continued long after to be conducted by foreign ships.

The withdrawal of the privileges of the foreign merchants in England had, however, an immediate effect in stimulating the home trade, as is proved by the fact, that in the year following the suppression of the foreign company, the English Merchant Adventurers shipped off for Flanders no less than 40,000 pieces of cloth. Myddelton entered into this new trade of cloth-making with great energy, and he prosecuted it with so much success, that in a speech delivered by him in the House of Commons on the proposed cloth patent, he stated that he and his partner then maintained several hundred families by that trade. He also seems to have taken part in the maritime adven-

---

[*] 'Tour in Wales,' vol. ii., p. 31. Ed. 1784.   [†] Williams's 'Ancient and Modern Denbigh,' p. 105.

tures of the period, most probably encouraged thereto by
his intimacy with Raleigh and other sea captains, including
his brother William, who had made profitable ventures on
the Spanish main. In short, Hugh Myddelton was regarded
as an eminently prosperous man.

At this stage of his affairs, when arrived at a comparatively advanced age, Myddelton took to himself a wife;
and the rank and fortune of the lady he married afford
some indication of the position he had by this time attained.
She was Miss Elizabeth Olmstead, the daughter and sole
heiress of John Olmstead of Ingatestone, Essex, with whom
the thriving goldsmith and merchant adventurer received a
considerable accession of property. That he had secured the
regard of his neighbours, and did not disdain to serve them
in the local offices to which they chose to elect him, is
apparent from the circumstance that he officiated for three
years as churchwarden for the parish of St. Matthew, to
which post he was appointed in the year 1598.

Myddelton continued to keep up a friendly connection
with his native town of Denbigh, and he seems to have
been mainly instrumental in obtaining for the borough its
charter of incorporation in the reign of Elizabeth. In return for this service the burgesses elected him their first
alderman, and in that capacity he signed the first by-laws
of the borough in 1597. On the back of the document are
some passages in his hand-writing, commencing with "Tafod
aur yngenau dedwydd" [A golden tongue is in the mouth of
the blessed], followed by other aphorisms, and concluding
with some expressions of regret at parting with his brethren,
the burgesses of Denbigh, whom he had specially visited on
the occasion. It would appear, from subsequent letters of his,
that about this time he temporarily resided in the town,—
most probably during an attempt which he made to sink for
coal in the neighbourhood, which turned out a total failure.

A few years later, Myddelton was appointed Recorder of
the borough, and in 1603 he was elected to represent it in
Parliament. In those days the office of representative was

not so much coveted as it is now, and boroughs remote from the metropolis were occasionally under the necessity of paying their members to induce them to serve.   It

Whitchurch, or St. Marcellus, Denbigh.

was, doubtless, an advantage to the burgesses of Denbigh that they had such a man to represent them as Hugh Myddelton, resident in London, and who was more-over an alderman and a benefactor of the town.   His two brothers—Thomas Myddelton, citizen and grocer, and Robert, citizen and skinner, of London—were members of the same Parliament, and we find Hugh and Robert fre-quently associated on committees of inquiry into matters connected with trade and finance.   Among the first com-mittees to which the brothers were appointed was one on the subject of a bill for explanation of the Statute of Sewers, and another for the bringing of a fresh stream of running water from the river of Lea, or Uxbridge, to the north parts of the city of London.   Thus the providing of a better supply of water to the inhabitants of the metropolis came very early under his notice, and doubtless had some influence in directing his future action on the subject.

At the same time the business in Bassishaw-street was not neglected, for, shortly after the arrival of King James in London, we find Myddelton supplying jewelry for Queen Anne, whose rage for finery of that sort was excessive. A warrant, in the State-Paper-office, orders 250l. to be paid to Hugh Myddelton, goldsmith, for a jewel given by James I. to the queen;* and it is probable that this connection with the Court introduced him to the notice of the king, and facilitated his approach to him when he afterwards had occasion to solicit His Majesty's assistance in bringing the New River works to completion.

The subject of water supply to the northern parts of the city was still under the consideration of parliamentary committees, of which Myddelton was invariably a member; and at length a bill passed into law, and the necessary powers were conferred. But no steps were taken to carry them into effect. The chief difficulty was not in passing the Act, but in finding the man to execute the work. A proposal made by one Captain Colthurst to bring a running stream from the counties of Hertford and Middlesex, was negatived by the Common Council in 1608. Fever and plague from time to time decimated the population, and the citizens of London seemed as far as ever from being supplied with pure water.

It was at this juncture that Hugh Myddelton stepped

---

* " 26th of February, 1604. To Hugh Middleton, Goldsmith, the sum of 250l. for a pendant of one diamond bestowed upon the Queen by His Majesty. By writ dated 9th day of January, 1604, 250l."— Extract from the 'Pell Records.' [The sum named would be equivalent to about 1000l. of our present money. The Queen's passion for jewels may be inferred from the circumstance stated by Dr. Steven in his 'Memoir of George Heriot,' the King's goldsmith (founder of Heriot's Hospital, Edinburgh), that during the ten years which immediately preceded the accession of King James to the throne of Great Britain, Heriot's bills for the Queen's jewels alone could not amount to less than 50,000l. sterling.]

forth and declared that if no one else would undertake it,
he would, and bring the water from Hertfordshire into
London. "The matter," quaintly observes Stow, "had
been well-mentioned though little minded, long debated but
never concluded, till courage and resolution lovingly shook
hands together, as it appears, in the soule of this no way to
be daunted, well-minded gentleman." When all others held
back—lord mayor, corporation, and citizens—Myddelton
took courage, and showed what one strong practical man,
borne forward by resolute will and purpose, can do. "The
dauntless Welshman," says Pennant, "stept forth and smote
the rock, and the waters flowed into the thirsting metropolis."

Myddelton's success in life seems to have been attributable
not less to his quick intelligence than to his laborious application and indomitable perseverance. He had, it is true,
failed in his project of finding coal at Denbigh; but the
practical knowledge which he acquired, during his attempt,
of the arts of mining and excavation, had disciplined his
skill and given him fertility of resources, as well as culti-
vated in him that power of grappling with difficulties, which
emboldened him to undertake this great work, more like
that of a Roman emperor than of a private London citizen.

The corporation were only too glad to transfer to him the
powers with which they had been invested by the legislature, together with the labour, the anxiety, the expense, and
the risk of carrying out an undertaking which they regarded as so gigantic. On the 28th of March, 1609, the corporation accordingly formally agreed to his proposal to bring
a supply of water from Amwell and Chadwell, in Hertfordshire, to Islington, as being "a thing of great consequence,
worthy of acceptation for the good of the city;" but subject
to his beginning the works within two months from the date
of their acceptance of his offer, and doing his best to finish
the same within four years. A regular indenture was drawn
up and executed between the parties on the 21st of April
following; and Myddelton began the works and "turned

the first sod" in the course of the following month, according to the agreement. The principal spring was at Chadwell, near Ware, and the operations commenced at that point. The second spring was at Amwell, near the same town; each being about twenty miles from London as the crow flies.

The general plan adopted by Myddelton in cutting the New River was to follow a contour line, as far as practicable, from the then level of the Chadwell Spring to the circular pond at Islington, subsequently called the New River Head. The stream originally presented a fall of about 2 inches in the mile, and its City end was at the level of about 82 feet above what is now known as Trinity high water mark. Where the fall of the ground was found inconveniently rapid, a stop-gate was introduced across the stream, penning from 3 to 4 feet perpendicularly, the water flowing over weirs down to the next level.

To accommodate the cut to the level of the ground as much as possible, numerous deviations were made, and the river was led along the sides of the hills, from which sufficient soil was excavated to form the lower bank of the

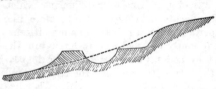

intended stream. Each valley was traversed on one side until it reached a point where it could be crossed; and there an embankment became necessary, in some cases of from 8 to 10 feet in height, along the top of which the water was conducted in a channel of the proper dimensions. In those places where the embankments were formed, provision had of course to be made for the passage of the surface waters from the west of the line of works into the river Lea, which forms the natural drain of the district. In some cases the drainage waters were conveyed under the New River in culverts, and in others over it by what were termed flashes. At each of the "flashes" there were extensive swamps, where the flood-waters were upheld to such a level as to enable them to pass over the flash, which consisted of a wooden trough, about twelve feet wide and three deep, extending across the river; and from these swamps, as well as from every other running stream, such apparatus was introduced as enabled the Company to avail themselves of the supply of water which they afforded, when required. Openings were also left in the banks for the passage of roads under the stream, the continuity of which was in such cases maintained either by arches or timber troughs lined with lead. One of these troughs, at Bush Hill, near Edmonton, was about 660 feet long, and 5 feet deep. A brick arch also formed part of this aqueduct, under which flowed a stream which had its source in Enfield Chase; the arch sustaining the trough and the road along its side. Another strong timber aqueduct, 460 feet long and 17 feet high, conducted the New River over the valley near where it entered the parish of Islington. This was

The Boarded River formerly at Bush Hill.

long known in the neighbourhood as "Myddelton's Boarded River." At Islington also there was a brick tunnel of considerable extent, and another at Newington. That at Islington averaged in section about 3 feet by 5, and appears to have been executed at different periods, in short lengths. Such were the principal works along the New River. Its original extent was much greater than it is at present, from its frequent windings along the high grounds for the purpose of avoiding heavy cuttings and embankments. Although the distance between London and Ware is only about 20 miles, the New River, as originally constructed, was not less than 38¾ miles in length.

Brick Arch under the New River, formerly near Bush Hill.

The works were no sooner begun than a swarm of opponents sprang up. The owners and occupiers of lands through which the New River was to be cut, strongly objected to it as most injurious to their interests. In a petition presented by them to Parliament, they alleged that their meadows

The map on the next page will enable the reader to trace the line of the New River works between Amwell, Chadwell, and London. The dotted lines indicate those parts of the old course which have since been superseded by more direct cuts, represented by the continuous black line. Where the loops have been detached from the present line of works, they are, in most instances, laid dry, and may be examined and measured correctly, as also the soil of which the banks were originally formed.

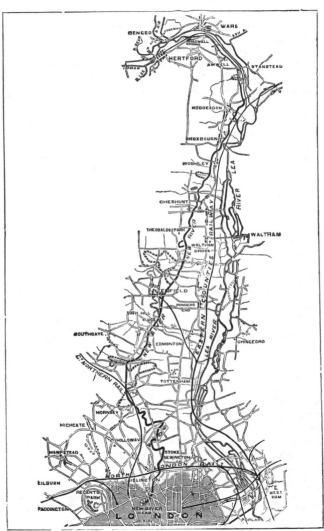

Map of the New River.

would be turned into "bogs and quagmires," and arable land
become "squallid ground;" that their farms would be
"mangled" and their fields cut up into quillets and "small
peeces;" that the "cut," which was no better than a deep
ditch, dangerous to men and cattle, would, upon "soden
raines," inundate the adjoining meadows and pastures, to the
utter ruin of many poor men; that the church would be
wronged in its tithe without remedy; that the highway
between London and Ware would be made impassable; and
that an infinity of evils would be perpetrated, and irretriev-
able injuries inflicted on themselves and their posterity.
The opponents also pointed out that the Mayor and corporation would have nothing to do with the business, but, by an
irrevocable act of the Common Council, had transferred
their powers of executing the works to Mr. Myddelton and
his heirs, "who doth the same for his own private benefit."

The agitation against the measure was next taken up in
Parliament. "Much ado there is in the House," writes
Mr. Beaulieu, on the 9th of May, 1610, to a friend in the
country, "about the work undertaken, and far advanced
already by Myddelton, of the cutting of a river and bringing it to London from ten or twelve miles off, through the
grounds of many men, who, for their particular interests,
so strongly oppose themselves to it, and are like (as it is
said) to overthrow it all." On the 20th of June following, a
Bill was introduced and committed to repeal the Act authorising the construction of the New River. A committee of ten
was appointed a few days after "to view" the river and to cer-
tify respecting the progress made with the works, doubtless
with the object of ascertaining what damage had actually
been done, or was likely to be done, to private property.
The committee were directed to make their report in the
next session; but as Parliament was prorogued in July, and
did not meet for four years, the subject is not again mentioned in the Journals of the House.

Worse than all, was the popular opposition which
Myddelton had to encounter. The pastor of Tottenham,

writing in 1631, speaks of the New River as " brought with
an ill wille from Ware to London." Stow, who was a
contemporary and enthusiastic admirer of Myddelton, says
bitterly, " If those enemies of all good endeavours, Danger,
Difficulty, Impossibillity, Detraction, Contempt, Scorn,
Derision, yea, and Desperate Despight, could have prevailed,
by their accursed and malevolent interposition, either before,
at the beginning, in the very birth of the proceeding, or in
the least stolne advantage of the whole prosecution, this
Worke, of so great worth, had never been accomplished."
Stow records that he rode down divers times to see the
progress made in cutting and constructing the New River,
and " diligently observed that admirable art, pains, and
industry were bestowed for the passage of it, by reason that
all grounds are not of a like nature, some being oozy and
very muddy, others again as stiff, craggy, and stony. The
depth of the trench," he adds, " in some places descended
full thirty feet, if not more, whereas in other places it
required a sprightful art again to mount it over a valley in
a trough, between a couple of hills, and the trough all the
while borne up by wooden arches, some of them fixed in the
ground very deep, and rising in height above twenty-three
feet."

It shortly became apparent to Myddelton that the time
originally fixed by the Common Council for the completion
of the works had been too short, and we accordingly find
him petitioning the Corporation for its extension. This was
granted him for five years more, on the ground of the opposition and difficulties which had been thrown in his way by
the occupiers and landowners along the line of the proposed
stream. It has usually been alleged that Myddelton fell
short of funds, and that the Corporation refused him the
necessary pecuniary assistance; but the Corporation records
do not bear out this statement, the only application apparently made by Myddelton being for an extension of time.
It has also been stated that he was opposed by the water-
carriers, and that they even stirred up the Corporation to

oppose the construction of the New River; but this statement seems to be equally without foundation. The principal obstacle which Myddelton had to encounter was
unquestionably the opposition of the landowners and
occupiers; and it was so obstinate that in his emergency he
was driven to apply to the King for assistance.

Though James I. may have been ridiculous and unkingly
in many respects, he nevertheless appears throughout his
reign to have exhibited a sensible desire to encourage the
industry and develop the resources of the kingdom he
governed. It was he who made the right royal declaration
with reference to the drowned lands in the Fens, that he
would not suffer the waters to retain their dominion over
the soil which skill and labour might reclaim for human
uses. He projected the drainage and reclamation of the
royal manor of Hatfield Chase, as well as the reclamation of
Sedgemoor and Malvern Chase; and when the landowners
in the Fens would take no steps to drain the Great Level,
he expressed the determination to become himself the sole
undertaker. And now, when Hugh Myddelton's admirable
project for supplying the citizens of London with water
threatened to break down by reason of the strong local
opposition offered to it, and while it was spoken of by many
with derision and contempt as an impracticable undertaking,
the same monarch came to his help, and while he rescued
Myddelton from heavy loss, it might be ruin, he enabled
him to prosecute his important enterprise to completion.

James had probably become interested in the works from
observing their progress at the point at which they passed
through the Royal Park at Theobalds, a little beyond
Enfield.* Theobalds was the favourite residence of the

---

* Theobalds, a singularly beautiful place, where Elizabeth held
counsel with Burleigh, James often
lived, and Charles played with his
children. The palace was ordered
to be pulled down by the Long Parliament, in spite of the commissioners' report that it was " an excellent building in very good repair:"
and, the materials having been sold
to the highest bidder, the proceeds
were divided amongst the soldiers
of Cromwell and Fairfax. The
materials alone realised not less
than 8275l. 11s.

King, where he frequently indulged in the pastime of hunting; and on passing the labourers occupied in cutting the New River, he would naturally make inquiries as to their progress. The undertaking was of a character so unusual, and so much of it passed directly through the King's domains, that he could not but be curious about it. Myddelton, having had dealings with His Majesty as a jeweller, seized the opportunity of making known his need of immediate help, otherwise the project must fall through. Several interviews took place between them at Theobalds and on the ground; and the result was that James determined to support the engineer with his effective help as King, and also with the help of the State purse, to enable the work to be carried out.

An agreement was accordingly entered into between the King and Myddelton, the original of which is deposited in the Rolls-office, and is a highly interesting document. It is contained on seven skins, and is very lengthy; but the following abstract will sufficiently show the nature of the arrangement between the parties. The Grant, as it is described, is under the Great Seal, and dated the 2nd of May, 1612. It is based upon certain articles of agreement, made between King James I. and Hugh Myddelton, "citizen and goldsmith of London," on the 5th of November preceding. It stipulates that His Majesty shall discharge a moiety of all necessary expenses for bringing the stream of water within "one mile of the city," as well as a moiety of the disbursements "already made" by Hugh Myddelton, upon the latter surrendering an account, and swearing to the truth of the same. In consideration of His Majesty's pecuniary assistance, Myddelton assigned to him a moiety of the interest in, and profits to arise from, the New River "for ever," with the exception of a small quill or pipe of water which the said Myddelton had granted, at the time of his agreement with the City, to the poor people inhabiting St. John-street and Aldersgate-street,—which exception His Majesty allowed.

One of the first benefits Myddelton derived from the arrangement was the repayment to him of one-half the expenditure which had been incurred to that time. It appears from the first certificate delivered to the Lord Treasurer, that the total expenditure to the end of the year 1612 had been 4485l. 18s. 11d., as attested by Hugh Myddelton, acting on his own behalf, and Miles Whitaores acting on behalf of the King. Further payments were made out of the Treasury for costs disbursed in executing the works; and it would appear from the public records that the total payments made out of the Royal Treasury on account of the New River works amounted to 8609l. 14s. 6d. As the books of the New River Company were accidentally destroyed by a fire many years ago, we are unable to test the accuracy of these figures by comparison with the financial records of the Company; but, taken in conjunction with other circumstances hereafter to be mentioned, the amount stated represents, with as near an approach to accuracy as can now be reached, one-half of the original cost of constructing the New River.

As the undertaking proceeded, with the powerful help of the King and the public Treasury, and as the great public uses of the New River began to be recognised, the voice of derision became gradually stilled, and congratulations began to rise up on all sides in view of the approaching completion of the bold enterprise. The scheme had ceased to be visionary, as it had at first appeared, for the water was already brought within a mile of Islington; all that was wanted to admit it to the reservoir being the completion of the tunnel near that place. At length that too was finished; and now King, Corporation, and citizens vied with each other in doing honour to the enterprising and public spirited Hugh Myddelton. The Corporation elected his brother Thomas Lord Mayor for the year; and on Michaelmas Day, 1613, the citizens assembled in great numbers to celebrate by a public pageant the admission of the New River water to the metropolis. The ceremony

took place at the new cistern at Islington, in the presence of
the Lord Mayor, Aldermen, Common Council, and a great concourse of spectators. A troop of some three score labourers in
green Monmouth caps, bearing spades and mattocks, or such
other implements as they had used in the construction of
the work, marched round the cistern to the martial music
of drums and trumpets, after which a metrical speech, composed by one Thomas Middleton, was read aloud, expressive
of the sentiments of the workmen. The following extract
may be given, as showing the character of the persons
employed on the undertaking :—

> First, here's the Overseer, this try'd man,
> An antient souldier and an artizan ;
> The Clearke ; next him the Mathematian ;
> The Maister of the Timber-worke takes place
> Next after these ; the Measurer in like case ;
> Bricklayer, and Enginer ; and after those
> The Borer, and the Pavier ; then it showes
> The Labourers next ; Keeper of Amwell Head ;
> The Walkers last ;—so all their names are read.
> Yet these but parcels of six hundred more,
> That, at one time, have been imploy'd before ;
> Yet these in sight and all the rest will say
> That all the weeke they had their Royall pay !

At the conclusion of the recitation the flood-gates were
thrown open, and the stream of pure water rushed into the
cistern amidst loud huzzas, the firing of mortars, the peal-
ing of bells, and the triumphant welcome of drums and
trumpets.[*]

---

[*] A large print was afterwards
published by G. Bickham, in commemoration of the event, entitled
'Sir Hugh Myddelton's Glory.' It
represents the scene of the ceremony,
the reservoir, with the stream rushing into it ; the Lord Mayor (Sir
John Swinnerton) on a white pal-
frey, pointing exultingly to Sir
Hugh ; the Recorder, Sir Henry
Montague, afterwards Lord Keeper
and Earl of Manchester, and by his
side the Lord Mayor elect, the pro-
jector's brother, Maister Thomas
Myddelton. Various figures gesti-
culating their admiration occupy
the foreground, whilst the foot of
the print is garnished with little
" chambers," or miniature mortars,
spontaneously exploding. There is
a copy of the original print in the
British Museum.

It is rather curious that James I. was afterwards nearly drowned in the New River which he had enabled Hugh Myddelton to complete. He had gone out one winter's day after dinner to ride in the park at Theobalds accompanied by his son Prince Charles; when, about three miles from the palace, his horse stumbled and fell, and the King was thrown into the river. It was slightly frozen over at the time, and the King's body disappeared under the ice, nothing but his boots remaining visible. Sir Richard Young rushed into his rescue, and dragged him out, when "there came much water out of his mouth and body." He was, however, able to ride back to Theobalds, where he got to bed and was soon well again. The King attributed his accident to the neglect of Sir Hugh and the Corporation of London in not taking measures to properly fence the river, and he did not readily forget it; for when the Lord Mayor, Sir Edward Barkham, accompanied by the Recorder, Sir Heneage Finch, attended the King at Greenwich, in June, 1622, to be knighted, James took occasion, in rather strong terms, to remind the Lord Mayor and his brethren of his recent mischance in "Myddelton's Water."

It is scarcely necessary to point out the great benefits conferred upon the inhabitants of London by the construction of the New River, which furnished them with an abundant and unremitting supply of pure water for domestic and other purposes. Along this new channel were poured into the city several millions of gallons daily; and the reservoirs at New River Head being, as before stated, at an elevation of 82 feet above the level of high water in the Thames, they were thus capable of supplying through pipes the basement stories of the greater number of houses then in the metropolis.

The pipes which were laid down in the first instance to convey the water to the inhabitants were made of *wood*, principally elm; and at one time the New River Company had wooden pipes laid down through the streets to the extent of about 400 miles! But the leakage was so great through the porousness of the material,—about one-fourth of the whole quantity of water supplied passing away by filtration,—and the decay of the pipes in ordinary weather was so rapid, besides being liable to burst during frosts, that they were ultimately abandoned when mechanical skill was sufficiently advanced to enable pipes of cast-iron to be substituted for them. For a long time, however, a strong prejudice existed against the use of water conveyed through pipes of any kind, and the cry of the water carriers long continued to be familiar to London ears, of " Any New River water here ! " " Fresh and fair New River water ! none of your pipe sludge ! "

" New River Water ! "
[After an Ancient Print.]

Among the many important uses to which the plentiful supply of New River water was put, was the extinction of fires, then both frequent and destructive, in consequence of the greater part of the old houses in London being built of wood. Stow particularly mentions the case of a fire which broke out in Broad Street, on the 12th November, 1623, in the house of Sir William Cockaigne, which speedily extended itself to several of the adjoining buildings. We are told by the chronicler, that " Sir Hugh Myddelton, upon the first knowledge thereof, caused all the sluices of the water-cisterne in the field to be left open, whereby there was plenty of water to quench the fire. The water " [of the New River], he continues, " hath done many like benefits in sundrie like former distresses."

We now proceed to follow the fortunes of Myddelton in connexion with the New River Company. The year after the public opening of the cistern at Islington, we find him a petitioner to the Corporation for a loan of 3000*l.*, for three years, at six per cent., which was granted him " in consideration of the benefit likely to accrue to the city from his New River; " his sureties being the Lord Mayor (Hayes), Mr. Robert Myddelton (his brother), and Mr. Robert Bateman. There is every reason to believe that Myddelton had involved himself in difficulties by locking up his capital in this costly undertaking; and that he was driven to solicit the loan to carry him through until he had been enabled to dispose of the greater part of his interest in the concern to other capitalists. This he seems to have done very shortly after the completion of the works. The capital was divided into seventy-two shares,* one-half of which belonged to Myddelton and the other half to the King, in consideration of the latter having borne one-half of the cost. Of the thirty-six shares owned by the former, as many as twenty-eight were conveyed by him to other persons; and that he realized a considerable sum by the sale is countenanced by the circumstance that we find him shortly after embarked in an undertaking hereafter to be described, requiring the command of a very large capital.

The shareholders were incorporated by letters patent on the 21st of June, 1619, under the title of " The Governors and Company of the New River brought from Chadwell and Amwell to London." † The government of the corporation was vested in the twenty-nine adventurers who held amongst them the thirty-six shares originally belonging to Myddelton, who had by that time reduced his holding

---

* In Pennant's 'London' it is stated that the original shares in the concern were 100*l.* each, whereas Entick makes them to have amounted to not less than 7000*l.* each! This is only another illustration of the hap-hazard statements put forth respecting Sir Hugh and his works. The original cost of the New River probably did not amount to more than 18,000*l.*, in which case the capital represented by each original share would be about 250*l.*

† On the occasion of its subsequent confirmation by parliament, Sir Edward Coke said : " This is a very good bill, and prevents one great mischief that hangs over the city. *Nimis potatio : frequens incendium.*"

to only two shares. At the first Court of proprietors, held on the 2nd of November, 1619, he was appointed Governor, and Robert Bateman Deputy-Governor of the Company. Sir Giles Mompesson was appointed, on behalf of the King, Surveyor of the profits of the New River, with authority to attend the meetings, inspect the accounts, &c., with a grant for such service of 200*l.* per annum out of the King's moiety of the profits of the said river. It was long, however, before there were any profits to be divided; for the cost of making repairs and improvements, and laying down wooden pipes, continued to be very great for many years; and the ingenious method of paying dividends out of capital, to keep up the price of shares and invite further speculation, had not yet been invented. In fact, no dividend whatever was paid until after the lapse of twenty years from the date of opening the New River at Islington; and the first dividend only amounted to 15*l.* 3*s.* 3*d.* a share. The next dividend of 3*l.* 4*s.* 2*d.* was paid three years later, in 1636; and as the concern seemed to offer no great prospect of improvement, and a further call on the proprietors was expected, Charles I., who required all his available means for other purposes, finally regranted his thirty-six "King's shares" to the Company, under his great seal, in consideration of a fee farm rent of 500*l.*, which is to this day paid by them yearly into the King's exchequer.

Notwithstanding this untoward commencement of the New River Company, it made great and rapid progress when its early commercial difficulties had been overcome; and after the year 1640 its prosperity steadily kept pace with the population and wealth of the metropolis. By the end of the seventeenth century the dividend paid was at the rate of about 200*l.* per share; at the end of the eighteenth century the dividend was above 500*l.* per share; and at the present date each share produces about 850*l.* a year. At only twenty years' purchase, the capital value of a single share at this day would be about 17,000*l.* But

most of the shares have in course of time, by alienation and
bequeathment, become very much subdivided; the pos-
sessors of two or more fractional parts of a share being
enabled, under a decree of Lord Chancellor Cowper, in
1711, to depute a person to represent them in the government of the Company.

Source of the New River at Chadwell, near Ware, with the Monumental Pedestal
in memory of Sir Hugh Myddelton.

CHAPTER IV.

Hugh Myddelton (*continued*) — His other Engineering and Mining Works — and Death.

Shortly after the completion of the New River, and the organization of the Company for the supply of water to the metropolis, we find Hugh Myddelton entering upon a new and formidable enterprise—that of enclosing a large tract of drowned land from the sea. The scene of his operations on this occasion was the eastern extremity of the Isle of Wight, at a place now marked on the maps as Brading Harbour. This harbour or haven consists of a tract of about eight hundred acres in extent. At low water it appears a wide mud flat, through the middle of which a small stream, called the Yar, winds its way from near the village of Brading, at the head of the haven, to the sea at its eastern extremity; whilst at high tide it forms a beautiful and apparently inland lake, embayed between hills of moderate elevation covered with trees, in many places down to the water's edge. At its seaward margin Bembridge Point stretches out as if to meet the promontory on the opposite shore, where stands the old tower of St. Helen's Church, now used as a sea-mark; and, as seen from most points, the bay seems to be completely landlocked.

The reclamation of so large a tract of land, apparently so conveniently situated for the purpose, had long been matter of speculation. It is not improbable that at some early period neither swamp nor lake existed at Brading Haven, but a green and fertile valley; for in the course of the works undertaken by Sir Hugh Myddelton for its recovery from the sea, a well, strongly cased with stone, was discovered near the middle of the haven, indicating the

View of Brading Haven, temporarily reclaimed by Sir Hugh Myddelton, as seen from
the Village of Brading.

existence of a population formerly settled on the soil.   The sea must, however, have burst in and destroyed the settlement, laying the whole area under water.

In King James's reign, when the inning of drowned lands began to receive an unusual degree of attention, the project of reclaiming Brading Haven was again revived; and in the year 1616 a grant was made of the drowned district to one John Gibb, the King reserving to himself a rental of 20l. per annum.   The owners of the adjoining lands contested the grant, claiming a prior right to the property in the haven, whatever its worth might be.   But the verdict of the Exchequer went against the landowners, and the right of the King to grant the area of the haven for the purpose of reclamation was maintained.   It appears that Gibb

Map of Brading Harbour.

sold his grant to one Sir Bevis Thelwall, a page of the King's bedchamber, who at once invited Hugh Myddelton to join him in undertaking the work; but Thelwall would not agree to pay Gibb anything until the enterprise had been found practicable. In 1620 we find that a correspondence was in progress as to "the composition to be made by the Solicitor-General

with Myddelton touching the draining of certain lands in
the Isle of Wight, and the bargain having been made
according to such directions as His Majesty hath given,
then to prepare the surrender, and thereupon such other
assurance for His Majesty as shall be requisite." *

A satisfactory arrangement having been made with the
King, Myddelton began the work of reclaiming the haven
in the course of the same year. He sent to Holland for
Dutch workmen familiar with such undertakings; and
from the manner in which he carried out his embankment,
it is obvious that he mainly followed the Dutch method of
reclamation, which, as we have already seen in the case of
the drainage of the Fens by Vermuyden, was not, in many
respects, well adapted for English practice. But it would
also appear, from a patent for draining land which he took
out in 1621, that he employed some invention of his own for
the purpose of facilitating the work. The introduction to
the grant of the patent runs as follows :—

" WHEREAS wee are given to vnderstand that our welbeloved
subiect Hugh Middleton, Citizen and Goldsmith of London, hath
to his very great charge maynteyned many strangers and others,
and bestowed much of his tyme to invent a new way, and by his
industrie, greate charge, paynes, and long experience, hath devised
and found out ' A NEW INVENÕON, SKILL, OR WAY FOR THE WYN-
NING AND DRAYNING OF MANY GROUNDℰ WHICH ARE DAYLIE
AND DESPERATELIE SURROUNDED WITHIN OUR KINGDOME OF ENG-
LAND AND DOMINION OF WALES,' and is now in very great hope to
bringe the same to good effect, the same not being heretofore
knowne, experimented, or vsed within our said realme or dominion,
whereby much benefitt, which as yet is lost, will certenly be brought
both to vs in particular and to our cõmon wealth in generall, and
hath offered to publish and practise his skill amongst our loving
subiectℰ . . . ., KNOWE YEE, that wee, tendring the weale
of this our kingdom and the benefitt of our subiectℰ , and out of our

---

* 'Domestic Calendar of State Papers.' Docquet, 13th August, 1620

princely care to nourish all art℮ , invencions, and studdies whereof there may be any necessary or ꝓffitable vse within our dominions, and out of our desire to cherish and encourage the industries and paynes of all other our loving subiect℮ in the like laudable indeavors, and to recompence the labors and expences of the said Hugh Middleton disbursed and to be susteyned as aforesaid, and for the good opinion wee have conceived of the said Hugh Middleton, for that worthy worke of his in bringing the New River to our cittie cf London, and his care and industrie in busines of like nature tending to the publicke good . . . . . doe give and graunt full, free, and absolute licence, libertie, power, and authoritie vnto the said Hughe Middleton, his deputies," &c. to use and practise the same during the terme of fowerteene years next ensuing the date hereof.

No description is given of the particular method adopted by Myddelton in forming his embankments. It would, however, appear that he proceeded by driving piles into the bottom of the Haven near Bembridge Point where it is

Entrance to Brading Harbour, from St. Helen's Old Tower.＊

* The above view represents the present state of the entrance to Brading Haven. A wide ridge of drifted sand lies across it, in front

about the narrowest, and thus formed a strong embankment at its junction with the sea, but unfortunately without making adequate provision for the egress of the inland waters.

A curious contemporary manuscript by Sir John Oglander is still extant, preserved amongst the archives of the Oglander family, who have held the adjoining lands from a period antecedent to the date of the Conquest, which we cannot do better than quote, as giving the most authentic account extant of the circumstances connected with the enclosing of Brading Haven by Hugh Myddelton. This manuscript says :—

" Brading Haven was begged first of all of King James by one Mr. John Gibb, being a groom of his bedchamber, and the man that King James trusted to carry the reprieve to Winchester for my Lord George Cobham and Sir Walter Rawleigh, when some of them were on the scaffold to be executed. This man was put on to beg it of King James by one Sir Bevis Thelwall, who was then one of the pages of the bedchamber. After he had begged it, Sir Bevis would give him nothing for it until the haven were cleared ; for the gentlemen of the island whose lands join to the haven challenged it as belonging unto them. King James was wonderful earnest in the business, both because it concerned his old servant, and also because it would be a leading case for the fens in Lincolnshire. After the verdict went in the Chequer against the gentlemen,

---

of the old bank raised by Myddelton, which extended from a point below the hill under " Mrs. Grant's house," a little to the westward of the village of Bembridge (seen on the opposite shore) to what are now called " The Boat Houses," situated towards the northern side of the haven, and behind the sand-ridge extending across the view. The black piles driven into the bottom of the haven in the process of embankment are still to be seen sticking up at low water; and only a few years since the old gates which served for a sluice were dug up near the Boat Houses. At the extremity of the sand-ridge there is a ferry across to the village of Bembridge, in front of which is the narrow entrance into the haven. There have been serious encroachments of the sea on that side of late years, and the channel has become much impeded : so much so that it has been feared that the navigation would be lost. The old church-tower of St. Helen's, faced with brick and whitewashed, on the right of the view, is still used as a sea-mark.

then Sir Bevis Thelwall would give nothing for it till he could see that it was feasible to be inned from the sea; whereupon one Sir Hugh Myddelton was called in to assist and undertake the work, and Dutchmen were brought out of the Low Countries, and they began to inn the haven about the 20th of December, 1620. Then, when it was taken in, King James compelled Thelwall and Myddelton to give John Gibb (who the King called 'Father') 2000l. Afterwards Sir Hugh Myddelton, like a crafty fox and subtle citizen, put it off wholly to Sir Bevis Thelwall, betwixt whom afterwards there was a great suit in the Chancery; but Sir Bevis did enjoy it some eight years, and bestowed much money in building of a barnhouse, mill, fencing of it, and in many other necessary works.

"But now let me tell you somewhat of Sir Bevis Thelwall and Sir Hugh Myddelton, and of the nature of the ground after it was inned, and the cause of the last breach. Sir Bevis was a gentleman's son in Wales, bound apprentice to a mercer in Cheapside, and afterwards executed that trade till King James came into England: then he gave up, and purchased to be one of the pages of the bedchamber, where, being an understanding man, and knowing how to handle the Scots, did in that infancy gain a fair estate by getting the Scots to beg for themselves that which he first found out for them, and then himself buying of them with ready money under half the value. He was a very bold fellow, and one that King James very well affected. Sir Hugh Myddelton was a goldsmith in London. This and other famous works brought him into the world, viz., his London waterwork, Brading Haven, and his mine in Wales.

"The nature of the ground, after it was inned, was not answerable to what was expected, for almost the moiety of it next to the sea was a light running sand, and of little worth. The best of it was down at the farther end next to Brading, my Marsh, and Knight's Tenement, in Bembridge. I account that there was 200 acres that might be worth 6s. 8d. the acre, and all the rest 2s. 6d. the acre. The total of the haven was 706 acres. Sir Hugh Myddelton, before he sold, tried all experiments in it: he sowed wheat, barley, oats, cabbage seed, and last of all rape seed, which proved best; but all the others came to nothing. The only inconvenience was in it that the sea brought in so much sand and ooze and seaweed that choked up the passage of the water to go out, insomuch as I am of opinion that if the sea had not broke in Sir

Bevis could hardly have kept it, for there would have been no current for the water to go out; for the eastern tide brought so much sand as the water was not of force to drive it away, so that in time it would have laid to the sea, or else the sea would have drowned the whole country. Therefore, in my opinion, it is not good meddling with a haven so near the main ocean.

"The country (I mean the common people) was very much against the inning of it, as out of their slender capacity thinking by a little fishing and fowling there would accrue more benefit than by pasturage; but this I am sure of, it caused, after the first three years, a great deal of more health in these parts than was ever before; and another thing is remarkable, that whereas we thought it would have improved our marshes, certainly they were the worse for it, and rotted sheep which before fatted there.

"The cause of the last breach was by reason of a wet time when the haven was full of water, and then a high spring tide, when both the waters met underneath in the loose sand. On the 8th of March, 1630, one Andrew Ripley that was put in earnest to look to Brading Haven by Sir Bevis Thelwall, came in post to my house in Newport to inform me that the sea had made a breach in the said haven near the easternmost end. I demanded of him what the charge might be to stop it out; he told me he thought 40s., whereupon I bid him go thither and get workmen against the next day morning, and some carts, and I would pay them their wages; but the sea the next day came so forcibly in that there was no meddling of it, for Ripley went up presently to London to Sir Bevis Thelwall himself, to have him come down and take some further course; but within four days after the sea had won so much on the haven, and made the breach so wide and deep, that on the 15th of March when I came thither to see it I knew not well what to judge of it, for whereas at the first 5l. would have stopped it out, now I think 200l. will not do it, and what will be the event of it time will tell. Sir Bevis on news of this breach came into the island on the 17th of March, 1630, and brought with him a letter from my Lord Conway to me and Sir Edward Dennies, desiring us to cause my Lady Worsley, on behalf of her son, to make up the breach which happened in her ground through their neglect. She returned us an answer that she thought that the law would not compel her unto it, and therefore desired to be excused, which answer we returned to my lord. What the event will be I know not, but it seemeth to me not reasonable that she should suffer for not complying with his request. If he had not inned the haven this accident could never have happened; there-fore he giving the cause, that she should apply the cure I understand not. But this I am sure, that Sir Bevis thinketh to recover of her and her son all his charges, which he now sweareth every way to be 2000*l*. For my part, I would wish no friend of mine to have any hand in the second inning of it. Truly all the better sort of the island were very sorry for Sir Bevis Thelwall, and the commoner sort were as glad as to say truly of Sir Bevis that he did the coun-try many good offices, and was ready at all times to do his best for the public and for everyone.

"Sir Hugh Myddelton took it first in, and it was proper for none but him, because he had a mine of silver in Wales to maintain it. It cost at the first taking of it in 4000*l*., then they gave 2000*l*. to Mr. John Gibb for it, who had begged it of King James; afterwards, in building the barn and dwelling-house, and water-mill, with the ditching and quick-setting, and making all the partitions, it could not have cost less than 200*l*. more: so in the total it stood them, from the time they began to take it in, until the 8th of March, a loss of 7000*l*."

It will thus be observed that the loss of this undertaking fell upon Thelwall, and not upon Myddelton, who sold out of the adventure long before the sea burst through the embankment. The date of conveyance of his rights in the reclaimed land to Sir Bevis Thelwall was the 4th September, 1624, nearly six years before the final ruin of the work. He had, therefore, got his capital out of the concern, most probably with his profit as contractor, and was thus free to embark in the important mining enter-prise in Wales, on which we find him next engaged.

Sir Hugh continued to maintain his Parliamentary con-nection with his native town of Denbigh, of which he was still the representative. We do not find that he took an active part in political questions. The name of his brother, Sir Thomas, frequently appears in the Parliamentary debates of the time, and he was throughout a strong oppo-nent of the Court party; but that of Sir Hugh only occurs in connection with commercial topics or schemes of internal improvement, on which he seems to have been consulted as an authority.

Sir Hugh's occasional visits to his constituents brought him into contact with Welsh families, and made him acquainted with the mining enterprises then on foot in different parts of Wales—so rich in ores of copper, lead, and iron. It appears that the Governor and Company of Mines Royal in Cardiganshire were incorporated in the year 1604, for the purpose of working the lead and silver mines of that county. The principal were those at Cwmsymlog and the Darren Hills, situated about midway, as the crow flies, between Aberystwith and the mountain of Plinlimmon, and at Tallybout, about midway between Aberystwith and the

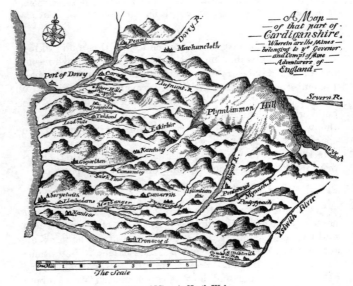

Chart of Mines in North Wales.

[From an old Print in the British Museum.]

estuary at the mouth of the River Dovey. They were all situated in the township of Skibery Coed, in the northern part of the county of Cardigan. For many years these mines (which were first opened out by the Romans) were worked by the Corporation of Mines Royal; but it does not appear that much success attended their operations. Mining was little understood then, and all kinds of pumping and lifting machinery were clumsy and inefficient. Although there was no want of ore, the mines were so drowned by water, that the metal could not well be got at and worked out.

Myddelton's spirit of enterprise was excited by the prospect of battling with the water and getting at the rich ore, and he had confidence that his mechanical ability would enable him to overcome the difficulties. The Company of Mines Royal were only too glad to get rid of their unprofitable undertaking, and they agreed to farm their mines to Sir Hugh at the rental of 400*l.* per annum. This was in the year 1617, some time after he had completed his New River works, but before he had begun the embankment of Brading Haven,—and Sir Bevis Thelwall was also a partner with him in this new venture. It took him some time to clear the mines of water, which he did by pumping-machines of his own contrivance; but at length sufficient ore was raised for testing, and it was found to contain a satisfactory proportion of silver. His mining adventure seems to have been attended with success, for we shortly afterwards find him sending considerable quantities of silver to the Royal Mint to be coined.

King James was so much gratified by the further proofs of Myddelton's skill and enterprise, displayed in his embankment of Brading Harbour and his successful mining operations in Wales, that he raised him to the dignity of a Baronet on the 19th of October, 1622; and the compliment was all the more marked by His Majesty directing that Sir Hugh should be discharged from the payment of the customary fees, amounting to 1095*l.*, and that the dignity

should be conferred upon him without any charge whatever.* The patent of baronetcy granted on the occasion
sets forth the "reasons and considerations" which induced
the King to confer the honour; and it may not be out of
place to remark, that though more eminent industrial
services have been rendered to the public by succeeding
engineers, there has been no such cordial or graceful recognition of them by any succeeding monarch. The patent
states that King James had made a baronet of Hugh
Myddelton, of London, goldsmith, for the following reasons
and considerations :—

"1. For bringing to the city of London, with excessive charge
and greater difficulty, a new cutt or river of fresh water, to the
great benefit and inestimable preservation thereof. 2. For gaining
a very great and spacious quantity of land in Brading Haven, in
the Isle of Wight, out of the bowells of the sea, and with bankes
and pyles and most strange defensible and chargeable mountains,
fortifying the same against the violence and fury of the waves.
3. For finding out, with a fortunate and prosperous skill, exceeding
industry, and noe small charge, in the county of Cardigan, a royal
and rych myne, from whence he hath extracted many silver plates
which have been coyned in the Tower of London for current money
of England." †

The King, however, did more than confer the title—he
added to it a solid benefit in confirming the lease made to
Sir Hugh by the Governor and Company of Mines Royal,
"as a recompense for his industry in bringing a new river
into London," waiving all claim to royalty upon the silver

---

* Sloane MS. (British Museum),
vol. ii. 4177. Also 'Calendar of
Domestic State Papers,' Oct. 19th,
1622: "Grant to Hugh Myddelton
of the rank of Baronet, granting
discharge of 1095*l.* due on being
made a Baronet." The usual statement is to the effect that Myddelton was knighted on the occasion
of the opening of the New River
in 1613. But this was not the

case, as it will be found from the
patent for draining land taken out
by him in 1621 (see *ante*, p. 86),
that he was then described as
simply "Citizen and Goldsmith, of
London." Nor is his name to be
found in any contemporary list of
King James's Knights.

† Harleian MS., No. 1507, **Art**
40. (British Museum.)

produced, although the Crown was entitled, according to the then interpretation of the law, to a payment on all gold and silver found in the lands of a subject; and it is certain that the lessee * who succeeded Sir Hugh did pay such royalty into the State Exchequer. It also appears from documents preserved amongst the State Papers, that large offers of royalty were actually made to the King at the very time that this handsome concession was granted to Sir Hugh.

The discovery of silver in the Welsh mountains doubtless caused much talk at the time, and, as in Australia and California now, there were many attempts made by lawless persons to encroach upon the diggings. On this, a royal proclamation was published, warning such persons against the consequences of their trespass, and orders were issued that summary proceedings should be taken against them. It appears that Sir Hugh and his partners continued to work the mines with profit for a period of about sixteen years, although it is stated that during most of that time, in consequence of the large quantity of water met with, little more than the upper surface could be got at. The water must, however, have been sufficiently kept under to enable so much ore eventually to be raised. Waller says an engine was employed at Cwmsymlog; and a tradition long existed among the neighbouring miners that there were two engines placed about the middle of the work. There were also several "levels" at Cwmsymlog, one of which is called to this day " Sir Hugh's Level."

The following rude cut, from Pettus' 'Fodinæ Regales,'

---

* Subsequent to 1636, Thomas Bushell (who purchased the lease) paid 1000l. per annum to the King; and some years after, in 1647, we find him agreeing to pay 2500l. per annum to the Parliament. As a curious fact, we may here add that, under date Die Sabbati, 14 August, 1641, Parliament granted an order or license to Thomas Bushell to dig turf on the King's wastes within the limits of Cardiganshire, for the purpose of smelting and refining the lead ores, &c., his predecessor (Myddelton) having used up almost all the wood growing in the neighbourhood of the mines.

may serve to give an idea of the manner in which the works of Cwmsymlog (facetiously styled by the author or his printer "Come-some-luck") were laid out:

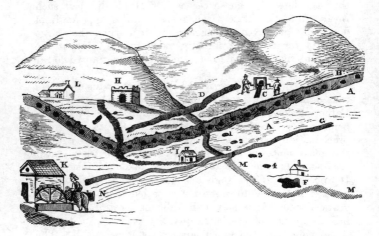

Plan of Myddelton's Silver Mining Works at Cwmsymlog.

A. The old works of Myddelton and Bushell.
B. The round holes are the shafts of the mine.
C. Windlace to wind up ore from the shafts.
D. A new vein.
E. Sir H. Myddelton's adit.
F. A new adit.
G. Adits to drain works.

H. Myddelton's decayed chapel.
I. Old stamping-house.
K. The smelting mills, supposed six miles from the hill.
L. Unwrought ground.
M. The brook that divides the hill.
N. The stream which drives the mills.

From a statement made by Bushell to Parliament of the results of the working subsequent to 1636, it appears that the lead alone was worth above 5000*l.* a year, to which there was to be added the value of the silver—Bushell alleging, in his petition to Charles I., deposited in the State Paper-office,* that Sir Hugh had brought "to the Minte theis 16 yeares of puer silver 100 poundes weekly."

Dated 22nd October, 1636. The prayer of Bushell's petition to Charles I. is, that His Majesty will ratify his agreement with Lady | Myddelton (by that time a widow) for the purchase of the residue of her lease.

A ton of the lead ore is said to have yielded about a hundred ounces of silver, and the yield at one time was such that Myddelton's profits were alleged by Bushell to have amounted to at least two thousand pounds a month. There is no doubt, therefore, that Myddelton realised considerable profits by the working of his Welsh mines, and that towards the close of his useful life he was an eminently prosperous man.*

Successful as he had been in his enterprise, he was ready to acknowledge the Giver of all Good in the matter. He took an early opportunity of presenting a votive cup, manufactured by himself out of the Welsh silver, to the corporation of Denbigh, and another to the head of his family at Gwaenynog, in its immediate neighbourhood, both of which are still preserved. On the latter is inscribed " Mentem non munus—Omnia a Deo—Hugh Myddelton."

While conducting the mining operations, Sir Hugh resided at Lodge, now called Lodge Park, in the immediate neighbourhood of the mines. The house was the property of Sir John Pryse, of Gogerddan, whose son Richard, afterwards created a baronet, was married to Myddelton's daughter Hester. The house stood on the top of a beautifully wooded hill, overlooking the estuary of the Dovey and the great bog of Gorsfochno, the view being bounded by picturesque hills on the one hand and by the sea on the other. Whilst residing here, on one of his visits to the mines, a letter reached him from his cousin, Sir John Wynn, of Gwydir, dated the 1st September, 1625, asking his assistance in an engineering project in which he was interested. This was the reclamation of the large sandy marshes, called Traeth-Mawr and Traeth-Bach, situated at

---

* In these rampant days of joint-stock enterprise, we are not surprised to observe that a scheme has been set on foot to work the long-abandoned silver and lead mines of Sir Hugh Myddelton. With the greatly improved mining machinery of modern times, it is not unreasonable to anticipate a considerable measure of success from such an undertaking.

the junction of the counties of Caernarvon and Merioneth, at the northern extremity of the bay of Cardigan. Sir John, after hailing his good cousin as "one of the great honours of the nation," congratulated him on the great work which he had performed in the Isle of Wight, and added, " I may say to you what the Jews said to Christ, We have heard of thy greate workes done abroad, doe now somewhat in thine own country." After describing the nature of the land proposed to be reclaimed, Sir John declares his willingness "to adventure a brace of hundred pounds to joyne with Sir Hugh in the worke," and concludes by urging him to take a ride to Traeth-Mawr, which was not above a day's journey from where Sir Hugh was residing, and afterwards to come on and see him at Gwydir House, which was at most only another day's journey or about twenty-five miles further to the north-west of Traeth-Mawr. The following was Sir Hugh's reply:—

" Honourable Sir,

"I have received your kind letter. Few are the things done by me; for which I give God the glory. It may please you to understand my first undertaking of public works was amongst my owne kindred, within less than a myle of the place where I hadd my first being, 24 or 25 years since, in seekinge of coales for the town of Denbighe.

" Touching the drowned lands near your lyvinge, there are many things considerable therein. Iff to be gayned, which will hardlie be performed without great stones, which was plentiful at the Weight [Isle of Wight], as well as wood, and great sums of money to be spent, not hundreds, but thousands ; * and first of all his Majesty's

---

* A long time passed before the attempt was made to reclaim the large tract of land at Traeth-Mawr ; but after the lapse of two centuries, it was undertaken by William Alexander Madocks, Esq., and accomplished in spite of many formidable difficulties. Two thousand acres of Penmorfa Marsh were first enclosed on the western side of the river, after which an embankment was constructed across the estuary, about a mile in length, by which 6000 additional acres were secured. The sums expended on the works are said to have exceeded 100,000l. ; but the expenditure has proved productive, and the principal part of the reclaimed land is now under cultivation. Tremadoc, or Madock's Town, and Port Madoc, are two thriving towns, built by the proprietor on the estate thus won from the sea.

interest must be got.   As for myself, I am grown into years, and full of business here at the mynes, the river at London, and other places, my weeklie charge being above 200*l.*; which maketh me verie unwillinge to undertake any other worke; and the least of theis, whether the drowned lands or mynes, requireth a whole man, with a large purse.   Noble sir, my desire is great to see you, which should draw me a farr longer waie; yet such are my occasions at this tyme here, for the settlinge of this great worke, that I can hardlie be spared one howr in a daie.   My wieff being also here, I cannot leave her in a strange place.   Yet my love to publique works, and desire to see you (if God permit), maie another tyme drawe me into those parts.   Soe with my heartie com̄endations I com̄it you and all your good desires to God.

" Your assured lovinge couzin to command,

" Lodge, Sept. 2nd, 1625."   " Hugh Myddelton.

At the date of this letter Sir Hugh was an old man of seventy, yet he still continued industriously to apply himself to business affairs.   Like most men with whom work has become a habit, he could not be idle, and active occupation seems to have been necessary to his happiness.   To the close of his life we find him engaged in correspondence on various subjects—on mining, draining, and general affairs. When in London he continued to occupy his house in Bassishaw-street, where the goldsmith business was carried on in his absence by his son William.   He also continued to maintain his pleasant country house at Bush Hill, near Edmonton, which he occupied when engaged on the engineering business of the New River, near to which it was conveniently situated.

At length all correspondence ceases, and the busy hand and head of the old man find rest in death.   Sir Hugh died on the 10th of December, 1631, at the advanced age of seventy-six.   In his will, which he made on the 21st November, three weeks before his death, when he was " sick in bodie " but " strong in mind," for which he praised God, he directed that he should be buried in the church of St. Matthew, Friday-street, where he had officiated as church-warden, and where six of his sons and five of his daughters

had been baptized. It had been his parish church, and was
hallowed in his memory by many associations of family
griefs as well as joys; for there he had buried several of his
children in early life, amongst others his two eldest-born
sons. The church of St. Matthew, however, has long since
ceased to exist, though its registers have been preserved: it
was destroyed in the great fire of 1666, and the monumental
record of Sir Hugh's last resting-place perished in the
common ruin.

The popular and oft-repeated story of Sir Hugh Myddelton having died in poverty and obscurity is only one of the
numerous fables which have accumulated about his memory.*
He left fair portions to all the children who survived him,
and an ample provision to his widow.† His eldest son and
heir, William, who succeeded to the baronetcy, inherited the
estate at Ruthin, and afterwards married the daughter of
Sir Thomas Harris, Baronet, of Shrewsbury. Elizabeth,
the daughter of Sir William, married John Grene, of
Enfield, clerk to the New River Company, and from her is

---

* The tradition still survives that
Sir Hugh retired in his old age to
the village of Kemberton, near
Shiffnal, Salop, where he lived in
great indigence under the assumed
name of Raymond, and that he was
there occasionally employed as a
street paviour! The parish register is said to contain an entry of his
burial on the 11th of March, 1702;
by which date Hugh Myddelton,
had he lived until then, would
have been about 150 years old!
The entry in the register was communicated by the rector of the
parish in 1809 to the 'Gentleman's
Magazine' (vol. lxxix., p. 795), but
it is scarcely necessary to point out
that it can have no reference whatever to the subject of this memoir.

† On the 24th June, 1632, Lady
Myddelton memorialised the Common Council of London with reference to the loan of 3000l. advanced

to Sir Hugh, which does not seem
to have been repaid; and more than
two years later, on the 10th Oct.,
1634, we find the Corporation allowed 1000l. of the amount, in consideration of the public benefit conferred on the city by Sir Hugh
through the formation of the New
River, and for the losses alleged to
have been sustained by him through
breaches in the water-pipes on the
occasion of divers great fires, as
well as for the "present comfort"
of Lady Myddelton. It is to be inferred that the balance of the loan
of 3000l. was then repaid. Lady
Myddelton died at Bush Hill on
the 19th July, 1643, aged sixty-
three, and was interred in the
chancel of Edmonton Church, Middlesex. On her monumental tablet it
is stated that she was "the mother
of fifteen children."

lineally descended the Rev. Henry Thomas Ellacombe, M.A., rector of Clyst St. George, Devon, who still holds two shares in the New River Company, as trustee for the surviving descendants of Myddelton in his family.   Sir Hugh left to his two other sons, Henry and Simon,* besides what he had already given them, one share each in the New River Company (after the death of his wife) and 400l. a-piece.   His five daughters seem to have been equally well provided for. Hester was left 900l., the remainder of her portion of 1900l.; Jane having already had the same portion on her marriage to Dr. Chamberlain, of London.   Elizabeth and Ann, like Henry and Simon, were left a share each in the New River Company and 500l. a-piece.   He bequeathed to his wife, Lady Myddelton, the house at Bush Hill, Edmonton, and the furniture in it, for use during her life, with remainder to his youngest son Simon and his heirs.   He also left her all the "chains, rings, jewels, pearls, bracelets, and gold buttons, which she hath in her custody and useth to wear at festivals, and the deep silver basin, spout pot, maudlin cup, and small bowl;" as well as "the keeping and wearing of the great jewel given to him by the Lord Mayor and Aldermen of London, and after her decease to such one of his sons as she may think most worthy to wear and enjoy it."   By the same will Lady Myddelton was authorised to dispose of her interest in the Cardiganshire mines for her own benefit; and it afterwards appears, from documents in the State Paper Office, that Thomas Bushell, "the great chymist," as he was called, purchased it for 400l. cash down, and 400l. per annum during the continuance of her grant, which had still twenty-five years to run after her husband's death.

Besides these bequeathments, and the gifts of land, money, and New River shares, which he had made to his other children during his lifetime, Sir Hugh left numerous other sums to relatives, friends, and clerks; for instance, to

---

Simon's son Hugh was created a Baronet, of Hackney, Middlesex, in 1681. He married Dorothy, the | daughter of Sir William Oglander, of Nunwell, Baronet.

Richard Newell and Howell Jones, 30*l.* each, "to the end that the former may continue his care in the works in the Mines Royal, and the latter in the New River water-works," where they were then respectively employed. He also left an annuity of 20*l.* to William Lewyn, who had been engaged in the New River undertaking from its commencement. Nor were his men and women servants neglected, for he bequeathed to each of them a gift of money, not forgetting "the boy in the kitchen," to whom he left forty shillings. He remembered also the poor of Henllan, near Denbigh, "the parish in which he was born," leaving to them 20*l.*; a similar sum to the poor of Denbigh, which he had represented in several successive Parliaments; and 5*l.* to the parish of Amwell, in Hertfordshire. To the Goldsmiths' Company, of which he had so long been a member, he bequeathed a share in the New River Company, for the benefit of the more necessitous brethren of that guild, "especially to such as shall be of his name, kindred, and county." *

Such was the life and such the end of Sir Hugh Myddelton, a man full of enterprise and resources, an energetic and untiring worker, a great conqueror of obstacles and difficulties, an honest and truly noble man, and one of the most distinguished benefactors the city of London has ever known.

---

* Several of the descendants of Sir Hugh Myddelton, when reduced in circumstances, obtained assistance from this fund. It has been stated, and often repeated, that Lady Myddelton, after her husband's death, became a pensioner of the Goldsmiths' Company, receiving from them 20*l.* a year. But this annuity was paid, not to the widow of the first Sir Hugh, but to the mother of the last Sir Hugh, more than a century later. The last who bore the title was an unworthy scion of this distinguished family. He could raise his mind no higher than the enjoyment of a rummer of ale; and towards the end of his life existed upon a pension granted him by the New River Company. The statements so often published (and which, on more than one occasion, have brought poor persons up to town from Wales to make inquiries) as to an annuity of 100*l.* said to have been left by Sir Hugh and unclaimed for a century, and of an advertisement calling upon his descendants to apply for the sum of 10,000*l.*, alleged to be lying for them at the Bank of England, are altogether unfounded. No such annuity has been left, no such sum has accrued, and no such advertisement has appeared.

CHAPTER V.

Captain Perry — Stoppage of Dagenham Breach.

Although the cutting of the New River involved a great deal of labour, and was attended with considerable cost, it was not a work that would now be regarded as of any importance in an engineering point of view. It was, nevertheless, one of the greatest undertakings of the kind that had at that time been attempted in England; and it is most probable that, but for the persevering energy of Myddelton and the powerful support of the King, the New River enterprise would have failed. As it was, a hundred years passed before another engineering work of equal importance was attempted, and then it was necessity, and not enterprise, that occasioned it.

We have, in a previous chapter, referred to the artificial embankment of the Thames, almost from Richmond to the sea, by which a large extent of fertile land is protected from inundation along both banks of the river. The banks first raised seemed to have been in many places of insufficient strength; and when a strong north-easterly wind blew down the North Sea, and the waters became pent up in that narrow part of it lying between the Belgian and the English coasts, —and especially when this occurred at a time of the highest spring tides,—the strength of the river embankments became severely tested throughout their entire length, and breaches often took place, occasioning destructive inundations.

Down to the end of the seventeenth century scarcely a season passed without some such accident occurring. There were frequent burstings of the banks on the south side between London Bridge and Greenwich, the district of Bermondsey, then green fields, being especially liable to be submerged. Commissions were appointed on such occasions with full powers to distrain for rates, and to impress labourers in order that the requisite repairs might at once be carried out. In some cases the waters for a long time held their ground, and refused to be driven back. Thus, in the reign of Henry VIII., the marshes of Plumstead and Lesnes, now used as a practising ground by the Woolwich garrison, were completely drowned by the waters which had burst through Erith Breach, and for a long time all measures taken to reclaim them proved ineffectual. There were also frequent inundations of the Combe Marshes, lying on the east of the royal palace at Greenwich.

But the most destructive inundations occurred on the north bank of the Thames. Thus, in the year 1676, a serious breach took place at Limehouse, when many houses were swept away, and it was with the greatest difficulty that the waters could be banked out again. The wonder is, that sweeping, as the new current did, over the Isle of Dogs, in the direction of Wapping, and in the line of the present West India Docks, the channel of the river was not then permanently altered. But Deptford was already established as a royal dockyard, and probably the diversion of the river would have inflicted as much local injury, judging by comparison, as it unquestionably would do at the present day. The breach was accordingly stemmed, and the course of the river held in its ancient channel by Deptford and Greenwich. Another destructive inundation shortly after occurred through a breach made in the embankment of the West Thurrock Marshes, in what is called the Long Reach, nearly opposite Greenhithe, where the lands remained under water for seven years, and it was with much difficulty that the breach could be closed.

But the most destructive and obstinate of all the breaches was that made in the north bank a little to the south of the village of Dagenham, in Essex, by which the whole of the Dagenham and Havering Levels lay drowned at every tide. A similar breach had occurred in 1621, which

Vermuyden, the Dutch engineer, succeeded in stopping; and at the same time he embanked or "inned" the whole of Dagenham Creek, through which the little rivulet flowing past the village of that name found its way to the Thames. Across the mouth of this rivulet Vermuyden had erected a sluice, of the nature of a "clow," being a strong gate suspended by hinges, which opened to admit of the egress of the inland waters at low tide, and closed against the entrance of the Thames when the tide rose. It happened, however, that a heavy inland flood, and an unusually high spring tide, occurred simultaneously during the prevalence of a strong north-easterly wind, in the year 1707; when the united force of the waters meeting from both directions blew up the sluice, the repairs of which had been neglected, and in a very short time nearly the whole area of the above Levels was covered by the waters of the Thames.

At first the gap was so slight as to have been easily closed, being only from 14 to 16 feet wide. But no measures having been taken to stop it, the tide ran in and out for several years, every tide wearing the channel deeper, and rendering the stoppage of the breach more difficult. At length the channel was found upwards of 30 feet deep at low water, and about 100 feet wide, a lake more than a mile and a half in extent having by this time been formed inside the line of the river embankment. Above a thousand acres of rich lands were spoiled for all useful purposes, and by the scouring of the waters out and in at every tide, the soil of about a hundred and twenty acres was completely washed away. It was carried into the channel of the Thames, and formed a bank of about a mile in length, reaching half-way across the river. This state of things could not be allowed to continue, for the navigation of the stream was seriously interrupted by the obstruction, and there was no knowing where the mischief would stop.

Various futile attempts were made by the adjoining landowners to stem the breach. They filled old ships with chalk and stones, and had them scuttled and sunk in the deepest places, throwing in baskets of chalk and earth outside them, together with bundles of straw and hay to stop up the interstices; but when the full tide rose, it washed them away like so many chips, and the opening was again driven clean through. Then the expedient was tried of sinking into the hole gigantic boxes made expressly for the purpose, fitted tightly together, and filled with chalk. Power was obtained to lay an embargo on the cargoes of chalk and ballast contained in passing ships, for the purpose of filling these boxes, as well as damming up the gap; and as many as from ten to fifteen freights of chalk a day were thrown in, but still without effect.

One day when the tide was on the turn, the force of the water lifted one of the monster trunks sheer up from the bottom, when it toppled round, the lid opened, out fell the chalk, and, righting again, the immense box floated out into the stream and down the river. One of the landowners interested in the stoppage ran along the bank, and shouted out at the top of his voice, "Stop her! stop her!" But the unwieldy object being under no guidance was carried down stream towards the shipping lying at Gravesend, where its unusual appearance, standing so high out of the water, excited great alarm amongst the sailors. The empty trunk, however, floated safely past, down the river, until it reached the Nore, where it stranded upon a sandbank.

The Government next lent the undertakers an old royal ship called the *Lion*, for the purpose of being sunk in the breach, which was done, with two other ships; but the *Lion* was broken in pieces by a single tide, and at the very next ebb not a vestige of her was to be seen. No matter what was sunk, the force of the water at high tide bored through underneath the obstacle, and only served to deepen the breach. After the destruction of the *Lion*, the channel was found deepened to 50 feet at low water, at the very place where she had been sunk.

All this had been but tinkering at the breach, and

every measure that had been adopted merely proved the incompetency of the undertakers. The obstruction to the navigation through the deposit of earth and sand in the river being still on the increase, an Act was passed in 1714, after the bank had been open for a period of seven years, giving powers for its repair at the public expense. But it is an indication of the very low state of engineering ability in the kingdom at the time, that several more years passed before the measures taken with this object were crowned with success, and the opening was only closed after a fresh succession of failures.

The works were first let to one Boswell, a contractor. He proceeded very much after the method which had already failed, sinking two rows of caissons or chests across the breach, but provided with sluices for the purpose of shutting off the inroads of the tide. All his contrivances, however, failed to make the opening watertight; and his chests were blown up again and again. Then he tried pontoons of ships, which he loaded and sunk in the opening; but the force of the tide, as before, rushed under and around them, and broke them all to pieces, the only result being to make the gap in the bank considerably wider and deeper than he found it. Boswell at length abandoned all further attempts to close it, after suffering a heavy loss; and the engineering skill of England seemed likely to be completely baffled by this hole in a river's bank.

The competent man was, however, at length found in Captain Perry, who had just returned from Russia, where, having been able to find no suitable employment for his abilities in his own country, he had for some time been employed by the Czar Peter in carrying on extensive engineering works.

John Perry was born at Rodborough, in Gloucestershire, in 1669, and spent the early part of his life at sea. In 1693 we find him a lieutenant on board the royal ship the *Montague*. The vessel having put into harbour at Portsmouth to be refitted, Perry is said to have displayed considerable mechanical skill in contriving an engine for throwing out a large quantity of water from deep sluices (probably for purposes of dry docking) in a very short space of time. The *Montague* having been repaired, went to sea, and was shortly after lost. As the English navy had suffered greatly during the same year, partly by mismanagement, and partly by treachery, the Government was in a very bad temper, and Perry was tried for alleged misconduct. The result was, that he was sentenced to pay a fine of 1000*l.*, and to undergo ten years' imprisonment in the Marshalsea.

This sentence must, however, have been subsequently mitigated, for we find him in 1695 publishing a " Regulation for Seamen," with a view to the more effectual manning of the English navy; and in 1698 the Marquis of Caermarthen and others recommended him to the notice of the Czar Peter, then resident in England, by whom he was invited to go out to Russia, to superintend the establishment of a royal fleet, and the execution of several gigantic works then contemplated for the purpose of opening up the resources of that empire. Perry was engaged by the Czar at a salary of 300*l.* a year, and shortly after accompanied him to Holland, thence proceeding to Moscow, to enter upon the business of his office.

One of the Czar's grand designs was to open up a system of inland navigation to connect his new city of St. Petersburg with the Caspian Sea, and also to place Moscow upon another line, by forming a canal between the Don and the Volga. In 1698 the works had been begun by one Colonel Breckell, a German officer in the Czar's service. But though a good military engineer, it turned out that he knew nothing of canal making; for the first sluice which he constructed was immediately blown up. The water, when let in, forced itself under the foundations of the work, and the six months' labour of several thousand workmen was destroyed in a night. The Colonel, having a due regard for his personal safety, at once fled the country in the disguise of a servant, and was never after heard of.

Captain Perry entered upon this luckless gentleman's office, and forthwith proceeded to survey the work he had begun, some seventy-five miles beyond Moscow. Perry had a vast number of labourers placed at his disposal, but they were altogether unskilled, and therefore comparatively useless. His orders were to have no fewer than 30,000 men at work, though he seldom had more than from 10,000 to 15,000; but one-twentieth the number of skilled labourers would have better served his purpose. He had many difficulties to contend with. The local nobility or boyars were strongly opposed to the undertaking, declaring it to be impossible; and their observation was, that God had made the rivers to flow one way, and it was presumption in man to think of attempting to turn them in another.

Shortly after the Czar had returned to his dominions, he got involved in war with Sweden, and was defeated by Charles XII. at the battle of Narva, in 1701. Athough the Don and Volga Canal was by this time half-dug, and many of the requisite sluices were finished, the Czar sent orders to Perry to let the works stand, and attend upon him immediately at St. Petersburg. Leaving one of his assistants to take charge of the work in hand, Perry waited upon his royal employer, who had a great new design on foot of an altogether different character. This was the formation of a royal dockyard on one of the southern rivers of Russia, where Peter contemplated building a fleet of war-ships, wherewith to act against the Turks in the Black Sea. Perry immediately entered upon the office to which he was appointed, of Comptroller of Russian Maritime Works, and proceeded to carry out the new project. The site of the Royal Dockyard was fixed at Veronize on the Don, where he was occupied for several years, with a vast number of workmen under him, in building a dockyard, with storehouses, ship-sheds, and workshops. He also laid down and superintended the construction of numerous vessels, one of them of eighty guns: the slips on which he built them are said to have been very ingeniously contrived.

The creation of this dockyard was far advanced when he received a fresh command to undertake the survey of a canal to connect St. Petersburg with the Volga, to enable provisions, timber, and building materials to flow freely to the capital from the interior of the empire. Perry surveyed three several routes, recommending the adoption of that through Lakes Ladoga and Onega; and the works were forthwith begun under his direction. Before they were completed, however, he had left Russia, never to return. During the whole of his stay in the kingdom he had been unable to get paid for his work. His applications for his stipulated salary were put off with excuses from year to year. Proceedings in the courts of law were out of the question in such a country; he could only dun the Czar and his ministers; and at length his arrears had become so great, and his necessities so urgent, that he could no longer endure his position, and threatened to quit the Czar's service. It came to his ears that the Czar had threatened on his part, that if he did, he would have Perry's head; and the engineer immediately took refuge at the house of the British minister, who shortly after contrived to get him conveyed safely out of the country, but without being paid. He returned to England in 1712, as poor as he had left it, though he had so largely contributed to create the navy of Russia, and to lay the foundations of its afterwards splendid system of inland navigation.

It will be remembered that all attempts made to stop the breach at Dagenham had thus far proved ineffectual; and it threatened to bid defiance to the engineering talent of England. Perry seemed to be one of those men who delight in difficult undertakings, and he no sooner heard of the work than he displayed an eager desire to enter upon it. He went to look at the breach shortly after his return, and gave in a tender with a plan for its repair; but on Boswell's being accepted, which was the lowest, he held back until that contractor had tried his best, and failed. The way was now clear for Perry, and again he offered to stop the

breach and execute the necessary works for the sum of
25,000*l*.* His offer was this time accepted, and operations
were begun early in 1715. The opening was now of great
width and depth, and a lake had been formed on the land
from 400 to 500 feet broad in some places, and extending
nearly 2 miles in length. Perry's plan of operations may
be briefly explained with the aid of his own map.

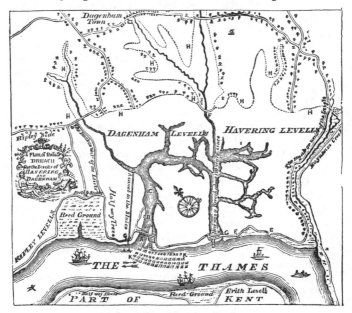

Perry's Plan of Dagenham Breach.

A.—The Dam whereby the Breach was stopped.
B.—The site of Boswell's works.
C.—The site of the Landowners' works.
D.—The site of Perry's Sluices.
E.—The site of Boswell's Sluices.
F.—A Dam and Sluice made for recovery of the Meadows shortly after the Breach had occurred.

G.—Small Sluice for drainage of the land waters.
HH.—The dotted line represents the extent of the inundation caused by the Breach.
I.—Places where stags' horns were dug up.
K.—Parallel lines, showing the depth at low water at every 60 yards distance from the shore.

---

* Boswell's price had been 16,300*l*., and he undertook to do the work
in fifteen months.

In the first place he sought to relieve the tremendous pressure of the waters against the breach at high tide, by making other openings in the bank through which they might more easily flow into and out of the inland lake, without having exclusively to pass through the gap which it was his object to stop. He accordingly had two openings, protected by strong sluices, made in the bank a little below the breach, and when these had been opened and were in action he proceeded to stop the breach itself. He began by driving in a row of strong timber piles across the channel; and they were dovetailed one into the other so as to render them almost impervious to water. The heads of the piles were not more than from eighteen inches to two feet above low water mark, so that in driving them little or no difficulty would be experienced from the current of ebb or flood. " Forty feet from this central row of sheeting piles, was constructed on each side, a sort of low coffer-dam-like structure, variously stated as 18 or 20 feet broad, formed of vertical piles and horizontal boarding, and filled with chalk, to prevent the toe of the future embankment from spreading. On the outside of these foot-wharfs, as Perry calls them, a wall of chalk rubble was made, as a further security. The dam itself was composed entirely of clayey earth, in layers about 3 feet in height, and scarcements or steps of about 7 feet; and in the course of its erection, care was taken always to shut the sluices already mentioned when, at each successive ebb-tide, the level of the back-water fell to the level of the top of the work in progress. In this way there was at no time a higher face for the water of the rising tide to flow over. In fact the unfinished embankment held in the water, over the land it was intended to lay dry, at a depth corresponding to its gradual progress, until finally, when the bank was above high-water line, it was discharged by the sluices, and never re-admitted." *

---

* Description of Perry's work by S. Downing, C.E., in *Practical Mechanics' Journal*, May, 1864.

Scarcely had Perry begun the work, and proceeded so far as to exhibit his general design, than Boswell, the former contractor, presented a petition to Parliament against the engineer being allowed to go on, alleging that his scheme was utterly impracticable. The work being of great importance, and executed at the public expense, a Parliamentary Committee was appointed, when Perry was called before them and examined fully as to the details. His answers were so explicit, and, on the whole, so satisfactory, that at the close of the examination one of the members thus spoke the sense of the Committee: — "You have answered us like an artist, and like a workman; and it is not only the scheme, but the man, that we recommend."

Perry was then allowed to proceed, and the work went steadily forward. About three hundred men were employed in stopping the breach, and it occupied them about five years to accomplish it. "Perry was proceeding steadily with the dam, which was constructed by successive scarcements about 7 feet broad and 3 feet high; these being supported by piles and planking on the side, and protected by layers of reeds on the top, had been able to resist the action of the tide when it came on. In this manner he was advancing to completion, when one of his assistants proposed to the parties who had advanced Captain Perry the necessary capital, to set all hands to work at neap tides, and form a narrow wall of earth, unprotected by reeds or planking, and build it so rapidly as to get it above the level of the springs before they should come on, and thus at once exclude the tides from the level. Unfortunately, the next spring-tide rose to an unexpected height under the influence of a storm from the north-west, and overtopped this narrow dam by about six inches, although Perry used the greatest energy, and heightened the wall of earth by piles and boarding set on edge on the top; but all in vain: the water poured over it, and in the course of *two hours* the whole dam was swept away, and the dovetailed piles laid bare. This accident was repaired in the winter months,

I

and in June, 1718, the tide was again turned out of the levels; but in September of the same year the dam gave way again, and this time with far greater injury to the work, as upwards of 100 feet of the dovetailed piles were torn up and carried away. In one place there was about 20 feet greater depth than before the work was begun. The third dam was completed on the 18th June, 1719, about fourteen years after the accident first occurred." Thus the opening was at length effectually stopped, and the water drained away by the sluices, leaving the extensive inland lake, which is to this day used by the Londoners as a place for fishing and aquatic recreation.*

Dagenham Lake.

A good idea of the formidable character of the embank

---

* It has recently been proposed to convert the lake into extensive docks, connected with London by the Tilbury Railway; and a Company has been formed for the purpose of carrying out the enterprise.

ments extending along the Thames may be obtained by a visit to this place.   Standing on the top of the bank, which is from 40 to 50 feet above the river level at low water,* we see on the one side the Thames, with its shipping passing and repassing, high above the inland level when the tide is up, with the still lake of Dagenham and the far extending flats on the other.   Looking from the lower level on these strong banks extending along the stream as far as the eye can reach, we can only see the masts of sailing ships and the funnels of large steamers leaving behind them long trails of murky smoke,—at once giving an idea of the gigantic traffic that flows along this great water highway, and the enormous labour which it has cost to bank up the lands and confine the river within its present artificial

The Thames from Dagenham Bank, looking up the River.

limits.   We do not exaggerate when we state that these formidable embankments, winding along the river side, up

---

* The banks themselves are from 17 to 25 feet high in the neighbourhood of Dagenham, and from 25 to 30 feet wide at the base.   The marks of the old breach are still easily traceable.

creeks and tributary streams, round islands and about marshes, from London to the mouth of the Thames, are not less than 300 miles in extent.

It is to be regretted that Perry gained nothing but fame by his great work. The expense of stopping the breach far exceeded his original estimate; he required more materials than he had calculated upon; and frequent strikes amongst his workmen for advances of wages greatly increased the total cost. These circumstances seem to have been taken into account by the Government in settling with the engineer, and a grant of 15,000*l.* was voted to him in consideration of his extra outlay. The landowners interested also made him a present of a sum of 1000*l.* But even then he was left a loser; and although the public were so largely benefited by the success of the work, which restored the navigation of the river, and enabled the adjoining proprietors again to reclaim for purposes of agriculture the drowned lands within the embankment, the engineer did not really receive a farthing's remuneration for his five years' anxiety and labour.

After this period Perry seems to have been employed on harbour works,—more particularly at Rye and Dover; but none of these were of great importance, the enterprise of the country being as yet dormant, and its available capital for public undertakings comparatively limited. It appears from the Corporation Records of Rye, that in 1724 he was appointed engineer to the proposed new harbour-works there. The port had become very much silted up, and for the purpose of restoring the navigation it was designed to cut a new channel, with two pier-heads, to form an entrance to the harbour. The plan further included a large stone sluice and draw-bridge, with gates, across the new channel, about a quarter of a mile within the pier-heads; a wharf constructed of timber along the two sides of the channel, up to the sluice; together with other well-designed improvements. But the works had scarcely been begun before the Commissioners displayed a strong disposition to job, one of them withdrawing for the purpose of supplying the stone and timber required for the new works at excessive prices, and others forming what was called "the family compact," or a secret arrangement for dividing the spoil amongst them. The plan of Perry was not fully carried out; and though the pier-heads and stone sluice were built, the most important part of the work, the cutting of the new channel, was only partly executed, when the undertaking was suspended for want of funds.

From that time forward, Perry's engineering ability was very much confined to making reports as to what things should be done, rather than in being employed to do them. In 1727 he published his "Proposals for Draining the Fens in Lincolnshire;" and he seems to have been employed there as well as in Hatfield Level, where "Perry's Drain" still marks one of his works. He was acting as engineer for the adventurers who undertook the drainage of Deeping Fen, in 1732, when he was taken ill and died at Spalding, in the sixty-third year of his age. He lies buried in the churchyard of that town; and the tombstone placed over his grave bears the following inscription :—

To the Memory of

IOHN PERRY Esq[r]; in 1693

Commander of His Maiesty King Will[ms]
Ship the Cignet; second Son of Sam[l] Perry
of Rodborough in Gloucestershire Gent & of
Sarah his Wife; Daughter of Sir Tho[s] Nott; K[r]
He was several Years Comptroller of the
Maritime works to Czar Peter in Russia &
on his Return home was Employed by y[e]
Parliament to stop Dagenham Breach which
he Effected and thereby Preserved the
Navigation of the River of Thames and
Rescued many Private Familys from Ruin
he after departed this Life in this Town &
was here Interred February 13; 1732 Aged
63 Years
This stone was placed over him by the
Order of William Perry of Penthurst in
Kent Esq[r] his Kindsman and Heir Male

CHAPTER VI.

The beginnings of Canal Navigation.

In the preceding memoirs of Vermuyden and Perry, we have found a vigorous contest carried on against the powers of water, the chief object of the engineers being to dam it back by embankments, or to drain it off by cuts and sluices; whilst in the case of Myddelton, on the other hand, we find his chief concern to have been to collect all the water within his reach, and lead it by conduit and aqueduct for the supply of the thirsting metropolis. The engineer whose history we are now about to relate dealt with water in like manner to Myddelton, but on a much larger scale; directing it into extensive artificial canals, for use as the means of communication between various towns and districts.

Down to the middle of last century, the trade and commerce of England were comparatively insignificant. This is sufficiently clear from the wretched state of our road and river communication about that time; for it is well understood that without the ready means of transporting commodities from place to place, either by land or water, commerce is impossible. But the roads of England were then about the worst in Europe, and usually impassable for vehicles during the greater part of the year.* Corn, wool, and such like articles, were sent to market on horses or bullocks' backs; and manure was carried to the field, and fuel conveyed from the forest or the bog, in the same way. The only coal used in the inland southern counties was carried on horseback in sacks for the supply of the blacksmiths' forges. The food of London was principally

---

* For a full account of Old Roads and Travelling in England, we must refer the reader to 'Lives of the Engineers,' vol. iii. pp. 1–73.

brought from the surrounding country in panniers. The little merchandise transported from place to place was mostly of a light description,—the cloths of the West of England, the buttons of Birmingham and Macclesfield, the baizes of Norwich, the cutlery of Sheffield, and the tapes, coatings, and fustians of Manchester.

Articles imported from abroad were in like manner conveyed inland by pack-horse or waggon; and it was then cheaper to bring most kinds of foreign wares from remote parts to London by sea than to convey them from the inland parts of England to London by road. Thus, two centuries since, the freight of merchandise from Lisbon to London was no greater than the land carriage of the same articles from Norwich to London; and from Amsterdam or Rotterdam the expense of conveyance was very much less. It cost from 7*l.* to 9*l.* to convey a ton of goods from Birmingham to London, and 13*l.* from Leeds to London. It will readily be understood that rates such as these were altogether prohibitory as regarded many of the articles now entering largely into the consumption of the great body of the people. Things now considered necessaries of life, in daily common use, were then regarded as luxuries, obtainable only by the rich. The manufacture of pottery was as yet of the rudest kind. Vessels of wood, of pewter, and even of leather, formed the principal part of the household and table utensils of genteel and opulent families; and we long continued to import our cloths, our linen, our glass, our " Delph " ware, our cutlery, our paper, and even our hats, from France, Spain, Germany, Flanders, and Holland. Indeed, so long as corn, fuel, wool, iron, and manufactured articles had to be transported on horseback, or in rude waggons dragged over still ruder roads by horses or oxen, it is clear that trade and commerce could make but little progress. The cost of transport of the raw materials required for food, manufactures, and domestic consumption, must necessarily have formed so large an item as to have in a great measure precluded their use; and before they could be made to enter largely into the general consumption, it was absolutely necessary that greater facilities should be provided for their transport.

England was not, however, like many other countries less favourably circumstanced, necessarily dependent solely upon roads for the means of transport, but possessed natural water communications, and the means of improving and extending them to an almost indefinite extent. She was provided with convenient natural havens situated on the margin of the world's great highway, the ocean, and had the advantage of fine tidal rivers, up which fleets of ships might be lifted at every tide into almost the heart of the land. Very little had as yet been done to take advantage of this great natural water power, and to extend navigation inland either by improving the rivers which might be made navigable, or by means of artificial canals, as had been done in Holland, France, and even Russia, by which those countries had in some parts been rendered in a great measure independent of roads.

It is true, public attention had from time to time been directed to the improvement of rivers and the cutting of canals, but excepting a few isolated attempts, little had been done towards carrying the numerous suggested plans in different parts of the country into effect. If we except some of the wider drains in the Fens, which were in certain cases made available for purposes of navigation, though to a very limited extent, the first canal was that constructed by John Trew, at Exeter, in 1566. In early times the tide carried vessels up to that city, but the Countess of Devon took the opportunity of revenging herself upon the citizens for some affront they had offered to her, by erecting a weir across the Exe at Topsham in 1284, which had the effect of closing the river to sea-going vessels. This continued until the reign of Henry VIII., when authority was granted by Parliament to cut a canal about three miles in length along the west side of the

river, from Exeter to Topsham. The work was executed
by Trew, and it is a curious circumstance that it contained
the first lock constructed in England,—though locks are
said to have been used in the Brenta in 1488, and were
shortly after adopted in the Milan canals. John Trew was
a native of Glamorganshire; and though he must have been
a man of skill and enterprise, like many other projectors of
improvements and benefactors of mankind, he seems to
have realised only loss and mortification by his work. In
consequence of an alleged failure on his part in carrying
out the agreement for executing the canal, the Mayor and
Chamber of the city disputed his claims, and he became
involved in ruinous litigation. In a letter written by him
to Lord Burleigh, in which he relates his suit against the
Chamber of Exeter, Trew draws a sad picture of the state
to which he was reduced. " The varyablenes of men,"
says he, " and the great injury done unto me, brought me
in such case that I wyshed my credetours sattisfyd and
I away from earth : what becom may of my poor wyf and
children, who lye in great mysery, for that I have spent
all." * He then proceeded to recount " the things whearin
God hath given (him) exsperyance ;" relating chiefly to
mining operations, and various branches of civil and even
military engineering. It is satisfactory to add that in 1573
the harassing suit was brought to a conclusion, and Trew
granted the Corporation a release on their agreeing to pay
him a sum of 224*l.*, and thirty pounds a year for life.†

In the reign of James I. several Acts of Parliament
were passed, giving powers to improve rivers, so as to
facilitate the passage of boats and barges carrying mer-

---

* Lansdowne MS. cvii. art. 73.

† The Lansdowne MS. xxxi.,
art. 74, sets forth certain " Reasons
against the proceedings of John
Trew in the works of Dover Haven,
1581." It appears from Lysons's
History of Dover, that Trew was
engaged as harbour-engineer at ten
shillings a-day wages ; but the
corporation, thinking that he was
inclined to protract the work on
which he was engaged, summarily
dismissed him. This is the last
that we hear of John Trew.

chandise. Thus, in 1623, Sir Hugh Myddelton was engaged upon a Committee on a bill then under consideration "for the making of the river of Thames navigable to Oxford." In the same year Taylor, the water poet, pointed out to the inhabitants of Salisbury that their city might be effectually relieved of its poor by having their river made navigable from thence to Christchurch. The progress of improvement, however, must have been slow; as urgent appeals, on the same subject, continued to be addressed to Parliament and the public for a century later.

In 1656 we find one Francis Mathew addressing Cromwell and his Parliament on the immense advantage of opening up a water-communication between London and Bristol. But he only proposed to make the rivers Isis and Avon navigable to their sources, and then either to connect their heads by means of a short sasse or canal of about three miles across the intervening ridge of country, or to form a fair stone causeway between the heads of the two rivers, across which horses or carts might carry produce between the one and the other. His object, it will be observed, was mainly the opening up of the existing rivers; "and not," he says, "to have the old channel of any river to be forsaken for a shorter passage." Mathew fully recognised the formidable character of his project, and considered it quite beyond the range of private enterprise, whether of individuals or of any corporation, to undertake it; but he ventured to think that it might not be too much for the power of the State to construct the three miles of canal and carry out the other improvements suggested by him, with a reasonable prospect of success. The scheme was, however, too bold for Mathew's time, and a century elapsed before another canal was made in England.

A few years later, in 1677, a curious work was published by Andrew Yarranton,* in which he pointed out

---

* 'England's Improvement by Sea and Land.'   London, 1677.

what the Dutch had accomplished by means of inland navigation, and what England ought to do as the best means of excelling the Dutch without fighting them. The main feature of his scheme was the improvement of our rivers so as to render them navigable and the inland country thus more readily accessible to commerce. For, in England, said he, there are large rivers well situated for trade, great woods, good wool and large beasts, with plenty of iron stone, and pit coals, with lands fit to bear flax, and with mines of tin and lead; and besides all these things in it, England has a good air. But to make these advantages available, the country, he held, must be opened up by navigation. First of all, he proposed that the Thames should be improved to Oxford, and connected with the Severn by the Avon to Bristol—these two rivers, he insisted, being the master rivers of England. When this has been done, says Mr. Yarranton, all the great and heavy carriage from Cheshire, all Wales, Shropshire, Staffordshire, and Bristol, will be carried to London and recarried back to the great towns, especially in the winter time, at half the rates they now pay, which will much promote and advance manufactures in the counties and places above named. "If I were a doctor," he says, "and could read a Lecture of the Circulation of the Blood, I should by that awaken all the City: For London is as the Heart is in the Body, and the great Rivers are as its Veins; let them be stopt, there will then be great danger either of death, or else such Veins will apply themselves to feed some other part of the Body, which it was not properly intended for: For I tell you, Trade will creep and steal away from any place, provided she may be better treated elsewhere." But he goes on—"I hear some say, You projected the making Navigable the River Stoure in Worcestershire: what is the reason it was not finished? I say it was my projection, and I will tell you the reason it was not finished. The River Stoure and some other Rivers were granted by an Act of Parliament to certain

Persons of Honour, and some progress was made in the work; but within a small while after the Act passed it was let fall again. But it being a brat of my own, I was not willing it should be Abortive; therefore I made offers to perfect it, leaving a third part of the Inheritance to me and my heirs for ever, and we came to an agreement. Upon which I fell on, and made it compleatly Navigable from Sturbridge to Kederminster; and carried down many hundred Tuns of Coales, and laid out near one thousand pounds, and then it was obstructed for Want of Money, which by Contract was to be paid."

There is no question that this "want of money" was the secret of the little progress made in the improvement of the internal communications of the country, as well as the cause of the backward state of industry generally. England was then possessed of little capital and less spirit, and hence the miserable poverty, starvation, and beggary which prevailed to a great extent amongst the lower classes of society at the time when Mr. Yarranton wrote, and which he so often refers to in the course of his book. For the same reason most of the early Acts of Parliament for the improvement of navigable rivers remained a dead letter: there was not money enough to carry them out, modest though the projects usually were. Among the few schemes which were actually carried out about the beginning of the eighteenth century, was the opening up of the navigation of the rivers Aire and Calder, in Yorkshire. Though a work of no great difficulty, Thoresby speaks of it in his diary as one of vast magnitude. It was, however, of much utility, and gave no little impetus to the trade of that important district.

It was, indeed, natural that the demand for improve ments in inland navigation should arise in those quarters where the communications were the most imperfect and where good communications were most needed, namely, in the manufacturing districts of the north of England. On the western side of the island Liverpool was then rising in

importance, and the necessity became urgent for opening
up its water communications with the interior. By the
assistance of the tide, vessels were enabled to reach as
high up the Mersey as Warrington; but there they were
stopped by the shallows, which it was necessary to remove to enable them to reach Manchester and the adjacent
districts.   Accordingly, in 1720, an Act was obtained
empowering certain persons to take steps to make navigable the rivers Mersey and Irwell from Liverpool to Manchester. This was effected by the usual contrivance of
wears, locks, and flushes, and a considerable improvement
in the navigation was thereby effected.   Acts were also
passed for the improvement of the Weaver navigation,
the Douglas navigation, and the Sankey navigation, all
in the same neighbourhood; and the works carried out
proved of much service to the district.

But these improvements, it will be observed, were principally confined to clearing out the channels of existing
rivers, and did not contemplate the making of new and
direct navigable cuts between important towns or districts.
It was not until about the middle of last century that
English enterprise was fairly awakened to the necessity
of carrying out a system of artificial canals throughout
the kingdom; and from the time when canals began to be
made, it will be found that the industry of the nation made
a sudden start forward. Abroad, monarchs had stimulated
like undertakings, and drawn largely on the public resources for the purpose of carrying them into effect; but
in England such projects are usually left to private enterprise, which follows rather than anticipates the public
wants. In the upshot, however, the English system, as it
may be termed — which is the outgrowth in a great
measure of individual energy—does not prove the least
efficient; for we shall find that the English canals, like
the English railways, were eventually executed with a
skill, despatch, and completeness, which imperial enterprise, backed by the resources of great states, was unable

to surpass or even to equal. How the first English canals were made, how they prospered, and how the system extended, will appear from the following biography of James Brindley, the father of canal engineering in England

F. Parsons   William Hall

James Brindley

(1716-72)

LIFE OF JAMES BRINDLEY.

CHAPTER I.

BRINDLEY'S EARLY YEARS.

IN the third year of the reign of George I., whilst the British Government were occupied in extinguishing the embers of the Jacobite rebellion which had occurred in the preceding year, the first English canal engineer was born in a remote hamlet in the High Peak of Derby.

The nearest town of any importance was Macclesfield, where a considerable number of persons were employed about the middle of last century in making wrought buttons in silk, mohair, and twist. The articles were sold throughout the country by pedestrian hawkers, most of whom had squatted on the waste lands called "The Flash," from a hamlet of that name situated between Buxton, Leek, and Macclesfield. They were notorious for their half-barbarous manners, and brutal pastimes. Such was the district, and such the population, in the neighbourhood of which our hero was born.

James Brindley first saw the light in a humble cottage standing about midway between the hamlet of Great Rocks, and that of Tunstead, in the liberty of Thornsett, some three miles to the north-east of Buxton. The house in which he was born, in the year 1716, has long since fallen to ruins—the Brindley family having been its last occupants. The walls stood for some time after the roof had fallen in, and at length the materials were removed to

I.   K build cowhouses; but in the middle of the ruin there grew up a young ash tree, forcing up one of the flags of the cottage floor. It looked so healthy and thriving a plant, that the labourer employed to remove the stones for the purpose of forming the pathway to the neighbouring farm-house, spared the seedling, and it grew up into the large and flourishing tree, six feet nine inches in girth, standing in the middle of the Croft, and now known as "Brindley's Tree." This ash tree is Nature's own memorial of the birth-place of the engineer, and it is the only one as yet erected in commemoration of his genius.

Brindley's Croft.*

Although the enclosure is called Brindley's Croft, this

* The site of the Croft is very elevated, and commands an extensive view as far as Topley Pike, between Bakewell and Buxton, at the top of what is called the Long Hill. Topley Pike is behind the spectator in looking at the Croft in the above aspect. The rising ground behind the ash tree is called Wormhill Common, though now enclosed. The old road from Buxton to Tideswell skirts the front of the rising ground.

name was only given to it of late years by its tenant, in memory of the engineer who was born there. The statement made in Mr. Henshall's memoir of Brindley,[*] to the effect that Brindley's father was the freehold owner of his croft, does not appear to have any foundation; as the present owner of the property, Dr. Fleming, informs us that it was purchased, about the beginning of the present century, from the heirs of the last of the Heywards, who became its owners in 1688. No such name as Brindley occurs in any of the title-deeds belonging to the property; and it is probable that the engineer's father was an under-tenant, and merely rented the old cottage in which our hero was born. There is no record of his birth, nor does the name of Brindley appear in the register of the parish of Wormhill, in which the cottage was situated; but registers in those days were very imperfectly kept, and part of that of Wormhill has been lost.

It is probable that Brindley's father maintained his family by the cultivation of his little croft, and that he was not much, if at all, above the rank of a cottier. It is indeed recorded of him that he was by no means a steady man, and was fonder of sport than of work. He went shooting and hunting, when he should have been labouring; and if there was a bull-running within twenty miles, he was sure to be there. The Bull Ring of the district lay less than three miles off, at the north end of Long Ridge Lane, which passed almost by his door; and of that place of popular resort Brindley's father was a regular frequenter. These associations led him into bad company, and very soon reduced him to poverty. He neglected his children, not only setting before them a bad example, but permitting them to grow up without education. Fortunately, Brindley's mother in a great measure supplied the father's shortcomings; she did what she could to teach them what she knew, though that was not much; but, perhaps more

---

[*] Kippis's 'Biographia Britannica,' Art. Brindley.

important still, she encouraged them in the formation of good habits by her own steady industry.*

The different members of the family, of whom James was the eldest, were thus under the necessity of going out to work at a very early age to provide for the family wants. James worked at any ordinary labourer's employment which offered until he was about seventeen years old. His mechanical bias had, however, early displayed itself, and he was especially clever with his knife, making models of mills, which he set to work in little mill-streams of his contrivance. It is said that one of the things in which he took most delight when a boy, was to visit a neighbouring grist-mill and examine the water-wheels, cog-wheels, drum-wheels, and other attached machinery, until he could carry away the details in his head; afterwards imitating the arrangements by means of his knife and such little bits of wood as he could obtain for the purpose. We can thus readily understand how he should have turned his thoughts in the direction in which we afterwards find him employed, and that, encouraged by his mother, he should have determined to bind himself, on the first opportunity that offered, to the business of a millwright.

The demands of trade were so small at the time, that Brindley had no great choice of masters; but at the village of Sutton, near Macclesfield, there lived one Abraham Bennett, a wheelwright and millwright, to whom young Brindley offered himself as apprentice; and in the year 1733, after a few weeks' trial, he became bound to that master for the term of seven years. Although the employment of millwrights was then of a very limited

---

* Brindley's father seems afterwards to have somewhat recovered himself; for we find him, in 1729, purchasing an undivided share of a small estate at Lowe Hill, within a mile of Leek, in Staffordshire, where he had before gone to settle; and he contrived to realise the remaining portion before his death, and to leave it to his son James. None of the Brindley family remained at Wormhill, and the name has disappeared in the district.

character, they obtained a great deal of valuable practical information whilst carrying on their business. The millwrights were as yet the only engineers. In the course of their trade they worked at the foot-lathe, the carpenter's bench, and the anvil, by turns; thus cultivating the faculties of observation and comparison, acquiring practical knowledge of the strength and qualities of materials, and dexterity in the handling of tools of many different kinds. In country places, where division of labour could not be carried so far as in the larger towns, the millwright was compelled to draw largely upon his own resources, and to devise expedients to meet pressing emergencies as they arose. Necessity thus made them dexterous, expert, and skilful in mechanical arrangements, more particularly those connected with mill-work, steam-engines, pumps, cranes, and such like. Hence millwrights in those early days were looked upon as a very important class of workmen. The nature of their business tended to render them self-reliant, and they prided themselves on the importance of their calling. On occasions of difficulty the millwright was invariably resorted to for help; and as the demand for mechanical skill arose, in course of the progress of manufacturing and agricultural industry, the men trained in millwrights' shops, such as Brindley, Meikle, Rennie, and Fairbairn, were borne up by the force of their practical skill and constructive genius into the highest rank of skilled and scientific engineering.

Brindley, however, only acquired his skill by slow degrees. Indeed, his master thought him slower than most lads, and even stupid. Bennett, like many well-paid master mechanics at that time, was of intemperate habits, and gave very little attention to his apprentice, leaving him to the tender mercies of his journeymen, who were for the most part a rough and drunken set. Much of the lad's time was occupied in running for beer, and when he sought for information he was often met with a rebuff. Skilled workmen were then very jealous of new hands, and those who were in any lucrative employment usually put their shoulders together to exclude outsiders. Brindley had thus to find out nearly everything for himself, and he only worked his way to dexterity through a succession of blunders.

He was frequently left in sole charge of the wheelwrights' shop—the men being absent at jobs in the country, and the master at the public-house, from which he could not easily be drawn. Hence, when customers called at the shop to get any urgent repairs done, the apprentice was under the necessity of doing them in the best way he could, and that often very badly. When the men came home and found tools blunted and timber spoiled, they abused Brindley and complained to the master of his bungling apprentice's handiwork, declaring him to be a mere "spoiler of wood." On one occasion, when Bennett and the journeymen were absent, he had to fit in the spokes of a cartwheel, and was so intent on completing his job that he did not find out that he had fitted them all in the wrong way until he had applied the gauge-stick. Not long after this occurrence, Brindley was left by himself in the shop for an entire week, working at a piece of common enough wheelwright's work, without any directions; and he made such a "mess" of it, that on the master's return he was so enraged, that he threatened, there and then, to cancel the indentures and send the young man back to farm-labourer's work, which Bennett declared was the only thing for which he was fit.

Brindley had now been two years at the business, and in his master's opinion had learnt next to nothing; though it shortly turned out that, notwithstanding the apprentice's many blunders, he had really groped his way to much valuable practical information on matters relating to his trade. Bennett's shop would have been a bad school for an ordinary youth, but it proved a profitable one for Brindley, who was anxious to learn, and determined to make a way for himself if he could not find one. He must have had a brave spirit to withstand the many difficulties he had to contend against, to learn dexterity through blunders, and success through defeats. But this is necessarily the case with all self-taught workmen; and Brindley was mainly self-taught, as we have seen, even in the details of the business to which he had bound himself apprentice.

In the autumn of 1735 a small silk-mill at Macclesfield, the property of Mr. Michael Daintry, sustained considerable injury from a fire at one of the gudgeons inside the mill, and Bennett was called upon to execute the necessary repairs. Whilst the men were employed at the shop in executing the new work, Brindley was sent to the mill to remove the damaged machinery, under the directions of Mr. James Milner, the superintendent of the factory. Milner had thus frequent occasion to enter into conversation with the young man, and was struck with the pertinence of his remarks as to the causes of the recent fire and the best means of avoiding similar accidents in future. He even applied to Bennett, his master, to permit the apprentice to assist in executing the repairs of certain parts of the work, which was reluctantly assented to. Bennett closely watched his "bungling apprentice," as he called him; but Brindley, encouraged by the superintendent of the mill, succeeded in satisfactorily executing his allotted portion of the repairs, not less to the surprise of his master than to the mortification of his men. Many years after, Brindley, in describing this first successful piece of mill-work which he had executed, observed, "I can yet remember the delight which I felt when my work was fixed and fitted complete; though I could not understand why my master and the other workmen, instead of being pleased, seemed to be dissatisfied with the insertion of every fresh part in its proper place."

The completion of the job was followed by the usual supper and drink at the only tavern in the town, then on Parsonage Green. Brindley's share in the work was a good deal ridiculed by the men when the drink began to operate; on which Mr. Milner, to whose intercession his participation in the work had been entirely attributable, interposed and said, "I will wager a gallon of the best ale in the house, that before the lad's apprenticeship is out he will be a cleverer workman than any here, whether master or man." We have not been informed whether the wager was accepted; but it was long remembered, and Brindley was so often taunted with it by the workmen, that he was not himself allowed to forget that it had been offered. Indeed, from that time forward, he zealously endeavoured so to apply himself as to justify the prediction, for it was nothing less, of his kind friend Mr. Milner; and before the end of his third year's apprenticeship his master was himself constrained to admit that Brindley was not the "fool" and the "blundering blockhead" which he and his men had so often called him.

Very much to the chagrin of the latter, and to the surprise of Bennett himself, the neighbouring millers, when sending for a workman to execute repairs in their machinery, would specially request that "the young man Brindley" should be sent them in preference to any other of the workmen. Some of them would even have the apprentice in preference to the master himself. At this Bennett was greatly surprised, and, quite unable to understand the mystery, he even went so far as to inquire of Brindley where he had obtained his knowledge of mill-work! Brindley could not tell; it "came natural-like;" but the whole secret consisted in Brindley working with his head as well as with his hands. The apprentice had already been found peculiarly expert in executing mill repairs, in the course of which he would frequently suggest alterations and improvements, more especially in the application of the water-power, which no one had before thought of, but which proved to be founded on correct principles, and worked to the millers' entire satisfaction. Bennett, on afterwards inspecting the gearing of one of the mills repaired by Brindley, found it so securely and substantially

fitted, that he even complained to him of his style of work.
" Jem," said he, " if thou goes on i' this foolish way o'
workin', there will be very little trade left to be done
when thou comes oot o' thy time : thou knaws firmness o'
wark's th' ruin o' trade." Brindley, however, gave no heed
whatever to the unprincipled suggestion, and considered it
the duty and the pride of the mechanic always to execute
the best possible work.

Among the other jobs which Brindley's master was
employed to execute about this time, was the machinery of
a new paper-mill proposed to be erected on the river Dane.
The arrangements were to be the same as those adopted in
the Smedley paper-mill on the Irk, and at Throstle-Nest,
on the Irwell, near Manchester ; and Bennett went over to
inspect the machinery at those places. But Brindley was
afterwards of opinion that he must have inspected the
taverns in Manchester much more closely than the paper-
mills in the neighbourhood ; for when he returned, the
practical information he brought with him proved almost a
blank. Nevertheless, Bennett could not let slip the oppor--
tunity of undertaking so lucrative a piece of employment in
his special line, and, ill-informed though he was, he set
his men to work upon the machinery of the proposed paper-
mill.

It very soon appeared that Bennett was altogether unfitted for the performance of the contract which he had
undertaken. The machinery, when made, would not fit ; it
would not work ; and, what with drink and what with
perplexity, Bennett soon got completely bewildered. Yet
to give up the job altogether would be to admit his own
incompetency as a mechanic, and must necessarily affect his
future employment as a millwright. He and his men,
therefore, continued distractedly to persevere in their
operations, but without the slightest appearance of satisfactory progress.

About this time an old hand, who happened to be passing
the place at which the men were at work, looked in upon

them and examined what they were about, as a mere
matter of curiosity. When he had done so, he went on to
the nearest public-house and uttered his sentiments on the
subject very freely. He declared that the job was a farce,
and that Abraham Bennett was only throwing his employer's money away. The statement of what the "experienced hand" had said, was repeated until it came to
the ears of young Brindley. Concerned for the honour of
his shop as well as for the credit of his master—though he
probably owed him no great obligation on the score either
of treatment or instruction—Brindley formed the immediate
resolution of attempting to master the difficulty so that the
work might be brought to a satisfactory completion.

At the end of the week's work Brindley left the mill
without saying a word of his intention to any one, and
instead of returning to his master's house, where he lodged,
he took the road for Manchester. Bennett was in a state
of great alarm lest he should have run away; for Brindley,
now in the fourth year of his apprenticeship, had reached
the age of twenty-one, and the master feared that, taking
advantage of his legal majority, he had left his service
never to return. A messenger was despatched in the course
of the evening to his mother's house; but he was not
there. Sunday came and passed—still no word of young
Brindley: he must have run away!

On Monday morning Bennett went to the paper-mill to
proceed with his fruitless work; and lo! the first person
he saw was Brindley, with his coat off, working away with
greater energy than ever. His disappearance was soon
explained. He had been to Smedley Mill to inspect the
machinery there with his own eyes, and clear up his master's difficulty. He had walked the twenty-five miles
thither on the Saturday night, and on the following Sunday
morning he had waited on Mr. Appleton, the proprietor of
the mill, and requested permission to inspect the machinery.
With an unusual degree of liberality Mr. Appleton gave the
required consent, and Brindley spent the whole of that

Sunday in the most minute inspection of the entire arrangements of the mill. He could not make notes, but he stored up the particulars carefully in his head; and believing that he had now thoroughly mastered the difficulty, he set out upon his return journey, and walked the twenty-five miles back to Macclesfield again.

Having given this proof of his determination, as he had already given of his skill in mechanics, Bennett was only too glad to give up the whole conduct of the contract thenceforth to his apprentice; Brindley assuring him that he should now have no difficulty in completing it to his satisfaction. No time was lost in revising the whole design; many parts of the work already fixed were rejected by Brindley, and removed; others, after his own design, were substituted; several entirely new improvements were added; and in the course of a few weeks the work was brought to a conclusion, within the stipulated time, to the satisfaction of the proprietors of the mill.

There was now no longer any question as to the extraordinary mechanical skill of Bennett's apprentice. The old man felt that he had been in a measure saved by young Brindley, and thenceforth, during the remainder of his apprenticeship, he left him in principal charge of the shop. For several years after, Brindley maintained his old master and his family in respectability and comfort; and when Bennett died, Brindley carried on the concern until the work in hand had been completed and the accounts wound up; after which he removed from Macclesfield to begin business on his own account at the town of Leek, in Staffordshire.

BRINDLEY'S NATIVE DISTRICT.

CHAPTER II.

BRINDLEY AS MASTER WHEELWRIGHT AND MILLWRIGHT.

BRINDLEY had now been nine years at his trade, seven as apprentice and two as journeyman; and he began business as a wheelwright at Leek at the age of twenty-six. He had no capital except his skill, and no influence except that which his character as a steady workman gave him. Leek was not a manufacturing place at the time when Brindley began business there in 1742. It was but a small market town, the only mills in the neighbourhood being a few grist-mills driven by the streamlets flowing into the waters of the Dane, the Churnet, and the Trent. These mills usually contained no more than a single pair of stones, and they were comparatively rude and primitive in their arrangement and construction.

Brindley at first obtained but a moderate share of employment. His work was more strongly done, and his charges were consequently higher, than was customary in the district; and the agricultural classes were as yet too poor to enable them to pay the prices of the best work. He gradually, however, acquired a position, and became known for his skill in improving old machinery or inventing such new mechanical arrangements as might be required for any special purpose. He was very careful to execute the jobs which were entrusted to him within the stipulated time, and he began to be spoken of as a thoroughly reliable workman. Thus his business gradually extended to other places at a distance from Leek, and more especially into the Staffordshire Pottery districts, about to rise into importance under the fostering energy of Josiah Wedgwood.

At first Brindley kept neither apprentices nor journeymen, but felled his own timber and cut it up himself, with such assistance as he could procure on the spot.   As his business increased he took in an apprentice, and then a journeyman, to carry on the work in the shop while he was absent; and he was often called to a considerable distance from home, more particularly for the purpose of being consulted about any new machinery that was proposed to be put up.   Nor did he confine himself to mill-work.   He was ready to undertake all sorts of machinery connected with the pumping of water, the draining of mines, the smelting of iron and copper, and the various mechanical arrangements connected with the manufactures rising into importance in the adjoining counties of Cheshire and Lancashire.   Whenever he was called upon in this way, he endeavoured to introduce improvements; and to such an extent did he carry this tendency, that he became generally known in the neighbourhood by the name of "The Schemer."

A number of Brindley's memoranda books* are still in existence, which show the varied nature of his employment during this early part of his career.   It appears from the entries made in them, that he was not only employed in repairing and fitting up silk-throwing mills at Macclesfield, all of which were then driven by water, but also in repairing corn-mills at Congleton, Newcastle-under-Lyne, and various other places, besides those in the immediate neighbourhood of Leek, where he lived.   We believe the pocket memoranda books, to which we refer, were the only records which Brindley kept of his early business transactions; the rest he carried in his memory, which by practice became remarkably retentive.   Whilst working as an apprentice at Macclesfield, he had taught himself the art of writing; but he never mastered it thoroughly, and to the end of his life he wrote with difficulty, and almost illegibly.   His spelling was also very bad; and what with the bad

---

* In the possession of Joseph Mayer, Esq., of Liverpool, who has kindly permitted the author to inspect the whole of his valuable manuscripts relating to Brindley, so curiously illustrative of his start in life as a working and consulting Engineer.

spelling and what with the hieroglyphics in which he wrote, it is sometimes very difficult to decypher the entries made by him from time to time in his books.

We find him frequently at Trentham. On one occasion he makes entry of a "Loog of Daal 20 foot long;" at another time he is fitting a pump for "Arle Gower," the Earl being one of Brindley's first patrons. The log of deal, it afterwards

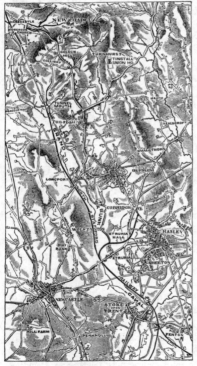

appears, was required for the flint-mill of a Mr. Tibots—"a mow [new?] invontion," as Brindley enters it in his book — of which more hereafter. On May 18, 1755, he enters "Big Tree to cut 1 day," and he seems to have felled the tree, and, some months after, to have cut it up himself, entering so many days at two shillings a day for the labour. When he had to travel some distance, he set down sixpence a day extra for expenses. Thus on one occasion he makes this entry: "For Mr. Kent corn mill of Codan looking out a shaft neer Broun Edge 1 day 0 : 2 : 6."

The Potteries District.

Between Leek and Trentham lay the then small pottery village of Burslem, which Brindley had frequent occasion to pass through in going to and from his jobs for the Earl. The earthenware then manufactured at Burslem was of a very inferior sort, consisting almost entirely of brown vessels; and the quantity turned out was so small that it was hawked about on the backs of the potters themselves, or sold by higglers, who carried it from village to village in the panniers of their donkeys. The brothers Elers, the Dutchmen, erected a potwork of an improved kind near Burslem, at the beginning of the century, in which they first practised the art of salt-glazing, brought by them from Holland.

The next improvement introduced was the use of powder of flints, used at first as a wash or dip, and afterwards mixed with tobacco-pipe clay, from which an improved ware was made, called "Flint potters." The merit of introducing this article is usually attributed to William Astbury, of Shelton, who, when on a journey to London, stopping at an inn at Dunstable, noticed the very soft and delicate nature of some burnt flint-stones when mixed with water (the hostler having used the powdered flint as a remedy for a disorder in his horses' eyes), and from thence he is said to have conceived the idea of applying it to the purposes of his trade. In first using the calcined flints, Mr. Astbury's practice was to have them pounded in an iron mortar until perfectly levigated; and being but sparingly used, this answered the demand for some time. But when the use of flint became more common, this tedious process would no longer suffice.

The brothers John and Thomas Wedgwood carried on the pottery business in a very small way, but were nevertheless hampered by an insufficient supply of flint powder, and it was found necessary to adopt some means of increasing it. In their emergency the potters called "The Schemer" to their aid; and hence we find him frequently occupied in erecting flint-mills, in Burslem and the neighbourhood, from that time forward. The success which attended his efforts brought Brindley not only fame, but business

It happened that, while thus occupied, Mr. John Edensor
Heathcote, owner of the Clifton estate near Manchester, became married to one of the daughters of Sir Nigel Gresley,
of Knypersley, in the neighbourhood of Burslem, and that
the marriage festivities were in progress, when the remarkable ingenuity of the young millwright of Leek was accidentally mentioned in the hearing of Mr. Heathcote one
day at dinner. The Manchester man, in the midst of
pleasure, did not forget business; and it occurred to him
that this ingenious mechanic might be of use in contriving some method for clearing his Clifton coal-mines of
the water by which they had so long been drowned. The
old methods of the gin-wheel and tub, and the chain-pump,
had been tried, but entirely failed to keep the water under:
if this Brindley could but do anything to help him in his
difficulty, he would employ him at once; at all events, he
would like to see the man.

Brindley was accordingly sent for, and the whole case
was laid before him. Mr. Heathcote described as minutely
as possible the nature of the locality, the direction in which
the strata lay, and exhibited a plan of the working of the
mines. Brindley was perfectly silent for a long time,
seemingly absorbed in a consideration of the difficulties to
be overcome; but at length his countenance brightened,
his eyes sparkled, and he briefly pointed out a method by
which he thought he should be enabled, at no great expense,
effectually to remedy the evil. His explanations were considered so satisfactory, that he was at once directed to
proceed to Clifton, with full powers to carry out his proposed plan of operations. This was, to call to his aid the
fall of the river Irwell, which formed one boundary of the
estate, and pump out the water from the pits by means of
the greater power of the water in the river.

With this object Brindley contrived and executed his
first tunnel, which he drove through the solid rock for a
distance of six hundred yards, and in this tunnel he led the
river on to the breast of an immense water-wheel fixed in a

chamber some thirty feet below the surface of the ground, from the lower end of which the water, after exercising its power, flowed away into the lower level of the Irwell. The expedient, though bold, was simple, and it proved effective. The machinery was found fully equal to the emergency; and in a very short time Brindley's wheel and pumps, working night and day, so cleared the mine of water as to enable the men to get the coal in places from which they had long been completely "drowned out."

We are not informed of the remuneration which the engineer received for carrying out this important work; but from the entries in his memorandum book it is probable that all he obtained was only his workman's wage of two shillings a day. Notwithstanding his ingenuity and hard-working energy, Brindley never seems, during the early part of his career, to have earned more than about one-third the wage of skilled mechanics in our own time; and from the insignificant sums charged by him for expenses, it is clear that he was satisfied to live in the fashion of an ordinary labourer. What modern engineers will receive ten guineas a day for doing, he, with his strong original mind, was quite content to do for two shillings. But eminent constructive skill seems to have been lightly appreciated in those days, if we may judge by the money value attached to it.* To this, however, it must be added, that

---

* The low remuneration paid for skilled labour of the same sort long before Brindley's time is worthy of passing notice. In 1544 John of Padua was paid only two shillings a-day as "devizour of His Majestie's works," in other words as Royal architect. Still later, Inigo Jones was paid only eight shillings and fourpence a-day as architect and surveyor of the Whitehall Banqueting House, and forty-six pounds a-year for house-rent, clerks, and incidental expenses; whilst Nicholas Stowe, the master mason, was allowed but four and tenpence a-day. When the Duchess of Marlborough was afterwards engaged in resisting the claims of one of her Blenheim surveyors, she told him indignantly "that Sir Christopher Wren, while employed upon Saint Paul's, was content to be dragged up to the top of the building three times a-week in a basket, at the great hazard of his life, for only 200*l.* a-year"—the actual amount of his salary as architect of that magnificent Cathedral. Brindley, however, fared worse still, and for

at the time of which we speak, the people of the country were comparatively poor—manufacturers as well as land-owners.

In Macclesfield and the neighbourhood, where the inventions of men such as Brindley have issued in so extraordinary a development of wealth, the operations of trade were as yet in their infancy, and had numerous obstructions and difficulties to contend against. Perhaps the greatest difficulty of all was the absence of those facilities for transport between one district and another, without which the existence of trade is simply impossible; but we shall shortly find Brindley also entering upon this great work of opening up the internal communications of the country, with an extraordinary degree of ability and success.

By the middle of last century, Macclesfield and the neighbouring towns were gradually rising out of the small button-trade, and aiming at greater things in the way of manufacture. In 1755 Mr. N. Pattison of London, Mr. John Clayton, and a few other gentlemen, entered into a partnership to build a new silk-mill at Congleton, in Cheshire, on a larger scale than had yet been attempted in that neighbourhood. Brindley was employed to execute the water-wheel and the commoner sort of mill-work about the building; but the smaller wheels and the more complex parts of the machinery, with which it was not supposed Brindley could be acquainted, were entrusted to a master joiner and millwright, named Johnson, who also superintended the progress of the whole work.

The superintendent required Brindley to work after his mere verbal directions, without the aid of any plan; and Brindley was not even allowed to inspect the models of the machinery required for the proposed mill. He thus worked at a great disadvantage, and the operations connected with

---

a long time does not seem to have risen above mere mechanic's pay, even whilst engaged in constructing the celebrated canal for the Duke of Bridgewater, which laid the foundation of so many gigantic fortunes.

the construction of the intended machinery were very
shortly found in a state of complete muddle. The proprie
tors had reason to suspect that their superintendent was not
equal to the enterprise which he had undertaken. At first
he endeavoured to assure them that all was going right;
but at last, after various efforts, he was obliged to confess
his incompetency and his inability to complete the work.

The proprietors, becoming alarmed, then sent for
Brindley and told him of their dilemma. "Would *he*
undertake to complete the works?" He asked to see the
model and plans which the superintendent engineer had
proposed to follow out. But on being applied to, the latter
positively refused to submit his designs to a common millwright, as he alleged Brindley to be. The proprietors were
almost in despair, and their only reliance now was on
Brindley's genius. "Tell me," he said, "what is the precise operation that you wish to perform, and I will endea-
vour to provide you with the requisite machinery for doing
it; but you must let me carry out the work in my own
way." To this they were only too glad to assent; and
having been furnished with the necessary powers, he forthwith set to work.

His intelligent observation of the process of manufacture
in the various mills he had inspected, his intimate practical
knowledge of machinery of all kinds then in use, and his
fertility of resources in matters of mechanical arrangement,
enabled him to perform even more than he had promised;
and he not only finished the mill to the complete satisfaction of its owners, but added a number of new and skilful
improvements in detail, which afterwards proved of the
greatest value. For instance, he adapted lifts to each
set of rollers and swifts, by means of which the silk
could be wound upon the bobbins equably, instead of
in wreaths as in other mills; and he so arranged the
shafting as to throw out of gear and stop either the whole
or any part of the machinery at will—an arrangement subsequently adopted in the throstle of the cotton-spinning

machine, and, though common enough now, then thought
perfectly marvellous. And, in order that the tooth-and-
pinion wheels should fit with perfect precision, he expressly
invented machinery for their manufacture—a thing that
had not before been attempted—all such wheels having,
until then, been cut by hand, at great labour and cost. By
means of this new machinery, as much work, and of a far
better description, could be cut in a day as had before occupied at least a fortnight. The result was, that the new
silk-mill, when finished, was found to be one of the most
complete and economical arrangements of manufacturing
machinery that had up to that time been erected in the
neighbourhood.

After the Congleton silk-mill had been completed, we
find Brindley engaged in erecting flint-mills in the Potteries,
of a more powerful and complete kind than any that had
before been tried, but which were rendered necessary by
the growing demands of the earthenware-manufacture.
One of the largest was that erected for Mr. Thomas
Baddely, at a place called Machins' of the Mill, near
Tunstall. We find these entries in Brindley's pocket-
book:—"March 15, 1757. With Mr. Badley to Matherso
about a now flint mill upon a windey day 1 day 3s. 6d.
March 19 draing a plann 1 day 2s. 6d. March 23 draing a
plann and to sat out the wheel race 1 day 4s."

This new mill was driven by water-power, and the wheel
both worked the pumping apparatus by which the adjoining
coal-mine was drained, and the stamping machinery for
pounding and grinding the flints. The wheel, which was
of considerable diameter, was fixed in a chamber below the
surface of the ground, and the water was conveyed to it
from the mill-pool through a small trough opening upon it
at its breast, which kept the paddle-boxes of the descending
part constantly filled, without any waste whatever, and
thus, by the rotation of the wheel, the pumps and stampers
were effectually worked. The main shaft was more than
two hundred yards from the mill; and to work the pumps

Brindley invented the slide rods, which were moved hori
zontally by a crank at the mill, and gave power to the
upright arm of a crank-lever, whose axis was at the angle,
and the lift at the other extremity. In course of time, as
improvements were introduced in the grinding of flints, the
stamping apparatus was detached from the machinery; but
this water-wheel continued its constant and useful operation
of pumping out the mines for full forty years after the
death of its inventor; and when it was at length broken
up, about the year 1812, the pump-trees, which consisted
of wooden staves firmly bound together with ashen hoops,
were found to be lined with cow-hides, the working
buckets being also covered with leather—a contrivance
of which the like, it is believed, has not before been
recorded.*

About the same time Brindley was requested by Mr.
John Wedgwood to erect a windmill for a similar purpose
on an elevated site adjoining the town of Burslem, called
The Jenkins; this being one of the first, if not the very
first, experiments made of the plan of grinding the calcined
flints in water, which in this case was pumped by the
action of the machinery from a well situated within the
mill itself. This invention, which was of considerable
importance, has by some been attributed to Brindley, whose
ingenious mind was ever ready to suggest improvements in
whatever process of manufacture came under his notice.
It was natural that he should closely watch the operation
of flint-grinding, having to construct and repair the greater
part of the machinery used in the process; and he could
not fail to notice the distressing consequences resulting
from inhaling the fine particles with which the air of the
flint-mills was laden. Hence the probability of his sug-
gesting that the flints should be ground in water, as calculated not only to prevent waste and preserve the purity of

* 'History of the Borough of Stoke-upon-Trent.' By John Ward.
1853. P. 164.

the air, but also to facilitate the operation of grinding,—a simple enough suggestion, but, as the result proved, a most valuable one.

With this object he invented an improved mill, which consisted of a large circular vat, about thirty inches deep, having a central step fixed in the bottom, to carry the axis of a vertical shaft. The moving power was applied to this shaft by a crown cog-wheel placed on the top. At the lower part of the shaft, at right angles to it, were four arms, upon which the grinding-stones were fixed, large blocks of stone of the same kind being likewise placed in the vat. These stones were a very hard silicious mineral, called " Chert," found in abundance in the neighbourhood of Bakewell, in Derbyshire. The broken flints being introduced to the vat and completely covered with water, the axis was made to revolve with great velocity, when the calcined flints were quickly reduced to an impalpable powder. This contrivance of Brindley's proved of great value to Wedgwood, and it was shortly after adopted throughout the Potteries, and continues in use to this day.

Being thus extensively occupied in the invention and erection of machinery driven by one power or another, it was natural that Brindley's attention should have been attracted to the use of steam power in manufacturing operations. Wind and water had heretofore been almost the exclusive agents employed for the purpose; but far-seeing philosophers and ingenious mechanics had for centuries been feeling their way towards the far greater power derived from the pent-up force of vaporised water; and engines had actually been contrived which rendered it likely that the problem would ere long be solved, and a motive agent invented, which should be easily controllable, and independent alike of wind, tides, and water-falls. Reserving for another place the history of the successive stages of this great invention, it will be sufficient for our present purpose merely to indicate, briefly, the direction of Brindley's labours in this important field.

It appears that Newcomen had as early as the year 1711 erected an atmospheric engine for the purpose of drawing water from a coal mine in the neighbourhood of Wolverhampton; and after considerable difficulties had been experienced in its construction and working, the engine was at length pronounced the most effective and economical that had yet been tried. Other engines of a similar kind were shortly after erected in the coal districts of the north of England, in the tin and copper mines of Cornwall, and in the lead mines of Cumberland, for the purpose of pumping water from the pits.

Brindley, like other contrivers of power, felt curious about this new invention, and proceeded to Wolverhampton to study one of Newcomen's engines erected there. He was greatly struck by its appearance, and, with the irrepressible instinct of the inventor, immediately set about contriving how it might be improved. He found the consumption of coal so great as to preclude its use excepting where coal was unusually abundant and cheap, as, for instance, at the mouth of a coal-pit, where the fuel it consumed was the produce and often the refuse of the mine itself; and he formed the opinion that unless the consumption of coal could be reduced, the extended use of the steam-engine was not practicable, by reason of its dearness, as compared with the power of horses, wind, or water.

With this idea in his head, he proceeded to contrive an improved engine, the main object of which was to ensure greater economy in fuel. In 1756 we find him erecting a steam-engine for one Mr. Broade, at Fenton Vivian, in Staffordshire, in which he adopted the expedient, afterwards tried by James Watt, of wooden cylinders made in the manner of coopers' ware, instead of cylinders of iron. He also substituted wood for iron in the chains which worked at the end of the beam. Like Watt, however, he was under the necessity of abandoning the wooden cylinders; but he surrounded

his metal cylinders with a wooden case, filling the intermediate space with wood-ashes; and by this means,
and using no more injection of cold water than was
necessary for the purpose of condensation, he succeeded in
reducing the waste of steam by almost one-half.

Whilst busy with Mr. Broade's engine, we find from
the entries in his pocket-book that Brindley occasionally
spent several days together at Coalbrookdale, in superintending the making of the boiler-plates, the pipes, and
other iron-work. Returning to Fenton Vivian, he proceeded with the erection of his engine-house and the
fitting of the machinery, whilst, during five days more,
he appears to have been occupied in making the hoops
for the cylinders. It takes him five days to get the
"great leavor fixed," thirty-nine days to put the boiler
together, and thirteen days to get the pit prepared; and
as he charges only workmen's wages for those days, we
infer that the greater part of the work was done by his
own hands. He even seems to have himself felled the
requisite timber for the work, as we infer from the entry
in his pocket-book of " falling big tree 3½ days."

The engine was at length ready after about a year's
work, and was set a-going in November, 1757, after which
we find these significant entries: "Bad louk [luck] five
days;" then, again, "Bad louk" for three days more; and,
after that, "Midlin louk;" and so on with "Midlin louk"
until the entries under that head come to an end. In the
spring of the following year we find him again striving
to get his "engon at woork," and it seems at length to
have been fairly started on the 19th of March, when we
have the entry "Engon at woork 3 days." There is then a
stoppage of four days, and again the engine works for seven
days more, with a sort of "loud cheer" in the words added
to the entry, of "driv a-Heyd!" Other intervals occur, until,
on the 16th of April, we have the words "at woor good
ordor 3 days," when the entries come to a sudden close.

The engine must certainly have given Brindley a great

deal of trouble, and almost driven him to despair, as we now know how very imperfect an engine with wooden hooped cylinders must have been; and we are not therefore surprised at the entry which he honestly makes in his pocket-book on the 21st of April, immediately after the one last mentioned, when the engine had, doubtless, a second time broken down, "to Run about a Drinking, 0 : 1 : 6." Perhaps he intended the entry to stand there as a warning against giving way to future despair; for he underlined the words, as if to mark them with unusual emphasis.*

Brindley did not remain long in this mood, but set to work upon the contrivance and erection of another engine upon a new and improved plan. What his plan was, may be learnt from the specification lodged in the Patent Office, on the 26th December, 1758, by "James Brindley, of Leek, in the county of Stafford, Millwright." † In the arrangement of this new steam-engine he provided that the boiler should be made of brick or stone arched over, and the stove over the fire-place of cast-iron, fixed within the boiler. The feeding-pipe for the boiler was to be made with a clack, opening and shutting by a float upon the surface of the water in the boiler, which would thus be self-feeding. The great chains for the segments at the extremity of the beams were of wood; and the pumps were also of wooden staves strongly hooped together.

---

* We find the following memorandum in Brindley's pocket-book, relating to the expense of working the engine in the year 1760 :—

Miss Clare Maria Broad⁵ fire engine at fentan vivian.

First yeer's work and repare night and day	£164
Do. torn back	025
Due for t⁶ first yeer	139
Due for the second yeer	102

† He describes it as "A Fire-Engine for Drawing Water out of Mines, or for Draining of Lands, or for Supplying of Cityes, Townes, or Gardens with Water, or which may be applicable to many other great and usefull Purposes, in a better and more effectual Manner than any Engine or Machine that hath hitherto been made or used for the like Purpose."—'Specifications of Patents,' No. 730.

Brindley, as a millwright, seems to have long retained his early predilection for wood, and to have preferred it to iron wherever its use was practicable. His plans were, however, subjected to modification and improvement from time to time, as experience suggested; and in the course of a few years, brick, stone, and wood were alike discarded in favour of iron; until, in 1763, we find Brindley erecting a steam-engine for the Walker Colliery, at Newcastle, wholly of iron, manufactured at Coalbrookdale, which was pronounced the most "complete and noble piece of ironwork" that had up to that time been produced.* But by this time Brindley's genius had been turned in another direction; the invention of the steam-engine being now safe in the hands of Watt, who was perseveringly occupied in bringing it to completion.

---

* Stuart's 'Anecdotes of Steam-Engines,' p. 626.

CHAPTER III.

The Duke of Bridgewater — Brindley employed as the Engineer of his Canal.

Very little had as yet been done to open up the inland navigation of England, beyond dredging and clearing out in a very imperfect manner the channels of some of the larger rivers, so as to admit of the passage of small barges. Several attempts had been made in Lancashire and Cheshire, as we have already shown, to open up the navigation of the Mersey and the Irwell from Liverpool to Manchester. There were similar projects for improving the Weaver from Frodsham, where it joins the Mersey, to Winford Bridge above Northwich; and the Douglas, from the Ribble to Wigan. About the same time like schemes were started in Yorkshire, with the object of opening up the navigation of the Aire and Calder to Leeds and Wakefield, and of the Don from Doncaster to near Sheffield.

One of the Acts passed by Parliament in 1737 is worthy of notice, as the forerunner of the Bridgewater Canal enterprise: we allude to the Act for making navigable the Worsley Brook to its junction with the river Irwell, near Manchester. A similar Act was obtained in 1755, for making navigable the Sankey Brook from the Mersey, about two miles below Warrington, to St. Helens, Gerrard Bridge, and Penny Bridge. In this case the canal was constructed separate from the brook, but alongside of it; and at several points locks were provided to adapt the canal to the level of the lands passed through.

The same year in which application was made to Parliament for powers to construct the Sankey Canal, the Corporation of Liverpool had under their consideration a much larger scheme—no less than a canal to unite the

Trent and the Mersey, and thus open a water-communication between the ports of Liverpool and Hull. It was proposed that the line should proceed by Chester, Stafford, Derby, and Nottingham. A survey was made, principally at the instance of Mr. Hardman, a public spirited merchant of Liverpool, and for many years one of its representatives in Parliament. Another survey was shortly after made at the instance of Earl Gower, afterwards Marquis of Stafford, and it was probably in making this survey that Brindley's attention was first directed to the business of canal engineering.

We find his first entry relating to the subject made on the 5th of February, 1758 — " novocion [navigation] 5 days;" the second, a little better spelt, on the 19th of the same month—" a bout the novogation 3 days;" and afterwards—" surveing the novogation from Long brigg to Kinges Milles 12 days ½." It does not, however, appear that the scheme made much progress, or that steps were taken at that time to bring the measure before Parliament; and Brindley continued to pursue his other employments, more especially the erection of " fire-engines " after his new patent. This continued until the following year, when we find him in close consultation with the Duke of Bridgewater relative to the construction of his proposed canal from Worsley to Manchester.

The early career of this distinguished nobleman was of a somewhat remarkable character. He was born in 1736, the fifth and youngest son of Scroop, third Earl and first Duke of Bridgewater, by Lady Rachel Russell. He lost his father when only five years old, and all his brothers died by the time that he had reached his twelfth year, at which early age he succeeded to the title of Duke of Bridgewater. He was a weak and sickly child, and his mental capacity was thought so defective, that steps were even in contemplation to set him aside in favour of the next heir to the title and estates. His mother seems almost entirely to have neglected him. In the first year of her widowhood she married Sir Richard Lyttleton, and from that time forward took the least possible notice of her boy.

The young Duke did not give much promise of surviving his consumptive brothers, and his mind was considered so incapable of improvement, that he was left in a great measure without either domestic guidance or intellectual discipline and culture. Horace Walpole writes to Mann in 1761: "You will be happy in Sir Richard Lyttleton and his Duchess; they are the best-humoured people in the world." But the good humour of this handsome couple was mostly displayed in the world of gay life, very little of it being reserved for home use. Possibly, however, it may have been even fortunate for the young Duke that he was left so much to himself, to profit by the wholesome neglect of special nurses and tutors, who are not always the most judicious in their bringing up of delicate children.

At seventeen, the young Duke's guardians, the Duke of Bedford and Lord Trentham, finding him still alive and likely to live, determined to send him abroad on his travels —the wisest thing they could have done. They selected for his tutor the celebrated traveller, Robert Wood, author of the well-known work on Troy, Baalbec, and Palmyra; afterwards appointed to the office of Under-Secretary of State by the Earl of Chatham. Wood was an accomplished scholar, a persevering traveller, and withal a man of good business qualities. His habits of intelligent observation could not fail to be of service to his pupil, and it is not unnatural to suppose that the great artificial watercourses and canals which they saw in the course of their travels had some effect in afterwards determining the latter to undertake the important works of a similar character by which his name became so famous. While passing through the south of France, the Duke was especially interested by his inspection of the Grand Canal of Languedoc, a magnificent work executed under great difficulties, and which had promoted in an extraordinary degree the prosperity of that part of the kingdom.* Proceeding into Italy, the Duke and his companion inspected all that was worthy of being seen there, including the picture galleries at Florence, Venice, and Rome. During their visit Mr. Wood sat to Mengs for his portrait, which still forms part of the Bridgewater collection. The Duke also purchased works of sculpture at Rome; but that he himself entertained no great enthusiasm for art is evident from the fact related by the late Earl of Ellesmere, that these works remained in their original packing-cases until after his death.†

Returned to England, he seems to have led the usual life of a gay young nobleman of the time, with plenty of money at his command. In 1756, when only twenty years old, he appears from the 'Racing Calendar' to have kept race-horses; occasionally riding them in matches himself. Though in after life a very bulky man, he was so light as a youth, that on one occasion Lord Ellesmere says a bet was jokingly offered that he would be blown off his horse. Dressed in a livery of blue silk and silver, with a jockey cap, he once rode a race against His Royal Highness the Duke of Cumberland, on the long terrace at the back of the wood in Trentham Park, the seat of his relative, Earl Gower. During His Royal Highness's visit, the large old green-house, since taken down, was hastily run up for the playing of skittles; and prison bars and other village games were instituted for the recreation of the guests. Those occupations of the Duke were varied by an occasional visit to his racing-stud at Newmarket, where he had a house for some time, and by the usual round of London gaieties during the season.

A young nobleman of tender age, moving freely in

---

* See Appendix, *Grand Canal of Languedoc*, and its execution by Riquet de Bonrepos.

† 'Essays in History, Biography, Geography, Engineering,' &c. By the late Earl of Ellesmere. London, 1858. P. 226.

circles where were to be seen some of the finest specimens
of female beauty in the world, could scarcely be expected
to pass heart-whole; and hence the occurrence of the
event in his London life which, singularly enough, is said
to have driven him in a great measure from society, and
induced him to devote himself to the construction of canals!
We find various allusions in the letters of the time to the
intended marriage of the young Duke of Bridgewater.
One rumour pointed to the only daughter and heiress of
Mr. Thomas Revell, formerly M.P. for Dover, as the object
of his choice. But it appears that the lady to whom he
became the most strongly attached was one of the Gun-
nings — the comparatively portionless daughters of an
Irish gentleman, who were then the reigning beauties at
Court. The object of the Duke's affection was Elizabeth,
the youngest daughter, and perhaps the most beautiful of
the three. She had been married to the fourth Duke of
Hamilton, in Keith's Chapel, Mayfair, in 1752, "with a
ring of the bed-curtain, half-an-hour after twelve at night,"*
but the Duke dying shortly after, she was now a gay and
beautiful widow, with many lovers in her train. In the
same year in which she had been clandestinely married to
the Duke of Hamilton, her eldest sister was married to the
sixth Earl of Coventry.

The Duke of Bridgewater paid his court to the young
widow, proposed, and was accepted. The arrangements
for the marriage were in progress, when certain rumours
reached his ear reflecting upon the character of Lady
Coventry, his intended bride's elder sister, who was certainly more fair than she was wise. Believing the reports,
he required the Duchess to desist from further intimacy
with her sister, a condition which her high spirit would
not brook, and, the Duke remaining firm, the match was
broken off. From that time forward he is said never

---

* 'Walpole to Mann' Feb. 27th, 1752.

to have addressed another woman in the language of gallantry.*

The Duchess of Hamilton, however, did not remain long a widow. In the course of a few months she was engaged to, and afterwards married, John Campbell, subsequently Duke of Argyll. Horace Walpole, writing of the affair to Marshal Conway, January 28th, 1759, says: " You and M. de Bareil do not exchange prisoners with half as much alacrity as Jack Campbell and the Duchess of Hamilton have exchanged hearts. . . . It is the prettiest match in the world since yours, and everybody likes it but the Duke of Bridgewater and Lord Conway. What an extraordinary fate is attached to these two women! Who could have believed that a Gunning would unite the two great houses of Campbell and Hamilton? For my part, I expect to see my Lady Coventry Queen of Prussia. I would not venture to marry either of them these thirty years, for fear of being shuffled out of the world prematurely to make room for the rest of their adventures."

The Duke, like a wise man, sought consolation for his disappointment by entering into active and useful occupation. Instead of retiring to his beautiful seat at Ashridge, we find him straightway proceeding to his estate at Worsley, on the borders of Chat Moss, in Lancashire, and conferring with John Gilbert, his land-steward, as to the practicability of cutting a canal by which the coals found upon his Worsley estate might be readily conveyed to market at Manchester.

Manchester and Liverpool at that time were improving towns, gradually rising in importance and increasing in population. The former place had long been noted for its manufacture of coarse cottons, or " coatings," made of wool,

---

* Chalmers, in his ' Biographical Dictionary,' vol. xiii., 94, gives another account of the rumoured cause of the Duke's subsequent antipathy to women; but the above statement of the late Earl of Ellesmere, confirmed as it is by certain passages in Walpole's Letters, is more likely to be the correct one.

Portrait of the Young Duke.

(1736–1803)

in imitation of the goods known on the Continent by that name. The Manchester people also made fustians, mixed stuffs, and small wares, amongst which leather-laces for women's bodices, shoe-ties, and points were the more important. But the operations of manufacture were still carried on in a clumsy way, entirely by hand. The wool was spun into yarn by means of the common spinning wheel, for the spinning-jenny had not yet been invented, and the yarn was woven into cloth by the common hand-loom. There was no whirr of engine-wheels then to be heard; for Watt's steam-engine had not yet come into existence. The air was free from smoke, except that which arose from household fires, and there was not a single factory-chimney in Manchester.

In 1724, Dr. Stukeley says Manchester contained no fewer than 2400 families, and that their trade was "incredibly large" in tapes, ticking, girth-webb, and fustians. In 1757 the united population of Manchester and Salford was only 20,000; * it is now, after the lapse of a century, 460,000! The Manchester manufacturer was then a very humble personage compared with his modern representative. He was part chapman, part weaver, and part merchant—working hard, living frugally, principally on oatmeal, and usually contriving to save a little money.

Dr. Aikin, writing in 1795, thus described the Manchester manufacturer in the first half of the eighteenth century: "An eminent manufacturer in that age," said he, "used to be in his warehouse before six in the morning, accompanied by his children and apprentices. At seven they all came in to breakfast, which consisted of one large dish of water-pottage, made of oatmeal, water, and a little salt, boiled thick, and poured into a dish. At the side was a pan or basin of milk, and the master and apprentices, each with a wooden spoon in his hand, without loss of time,

* Aikin's 'Description of the Country from Thirty to Forty | Miles round Manchester.' London, 1795.

dipped into the same dish, and thence into the milk-pan, and as soon as it was finished they all returned to their work." What a contrast to the "eminent manufacturer" of our own day!

As trade increased, its operations became more subdivided, and special classes and ranks began to spring into import-

The South West Prospect of Manchester and Salford

View of Manchester in 1740.
[Fac-simile of an Engraving of the period.]

ance.   The manufacturers sent out riders to take orders, and gangs of chapmen with pack-horses to distribute the goods and bring back wool, which they either used up themselves, or sold to makers of worsted yarn at Manchester, or to the clothiers of Rochdale, Saddleworth, or the West Riding of Yorkshire.   Mr. Walker, author of the 'Original,' left the following interesting reminiscence of the dealings of Manchester men with the inhabitants of the Fen districts:—"I have by tradition," said he, "the following particulars of the mode of carrying on the home trade by one of the principal merchants of Manchester, who was born at the commencement of the last century, and who realised a sufficient fortune to keep a carriage when not half a dozen were kept in the town by persons connected with business.   He sent the manufactures of the place into

Nottinghamshire, Lincolnshire, Cambridgeshire, and the intervening counties, and principally took in exchange feathers from Lincolnshire, and malt from Cambridgeshire and Nottinghamshire. All his commodities were conveyed on pack-horses, and he was from home the greater part of every year, performing his journeys entirely on horseback. His balances were received in guineas, and were carried with him in his saddle-bags. He was exposed to the vicissitudes of the weather, to great labour and fatigue, and to constant danger. In Lincolnshire he travelled chiefly along bridle-ways through fields where frequent gibbets warned him of his perils, and where flocks of wild fowl continually darkened the air. Business carried on in this manner required a combination of personal attention, courage, and physical strength, not to be hoped for in a deputy; and a merchant then led a much more severe and irksome life than a bagman afterwards, and still more than a traveller of the present day. In the earlier days of the merchant abovementioned, the wine merchant who supplied Manchester, resided at Preston, then always called Proud Preston, because exclusively inhabited by gentry. The wine was carried on horses, and a gallon was considered a large order. Men in business confined themselves generally to punch and ale, using wine only as a medicine, or on extraordinary occasions; so that a considerable tradesman somewhat injured his credit amongst his neighbours by being so extravagant as to send to a tavern for wine, to entertain a London customer." *

The roads out of Manchester in different directions, like those in most districts throughout the kingdom, were in a very neglected state, being for the most part altogether impracticable for waggons. Hence the use of pack-horses was an absolute necessity; and the roads were but ill-

---

* Thomas Walker: The Original, No. xi. Article entitled "Change in Commerce."

adapted even for them. Indeed, it was more difficult then to reach a village twenty miles out of Manchester than it is to make the journey from thence to London now. The only coach to London plied but every second day, and it was four days and a half in making the journey, there being a post only three times a week.* The roads in most districts of Lancashire were what were called "mill roads," along which a horse with a load of oats upon its back might proceed towards the mill where they were to be ground. There was no private carriage kept by any person in business in Manchester until the year 1758, when the first was set up by some specially luxurious individual. But wealth led to increase of expenditure, and Aikin mentions that there was "an evening club of the most opulent manufacturers, at which the expenses of each person were fixed at fourpence-halfpenny—fourpence for ale, and a halfpenny for tobacco." The progress of luxury was further aided by the holding of a dancing assembly once a week in a room situated about the middle of King Street, now a busy thoroughfare, the charge for admission to the nightly ball being half-a-crown the quarter. The ladies had their maids to wait for them with lanterns and pattens, and to conduct them home; "nor," adds Aikin, "was it unusual for their partners also to attend them."

The imperfect state of the communications leading to and from Manchester rendered it a matter of some difficulty at certain seasons to provide food for so large a population. In winter, when the roads were closed, the place was in the condition of a beleaguered town; and even in summer, the land about Manchester itself being comparatively sterile, the place was badly supplied with fruit, vegetables, and potatoes, which, being brought from con-

---

* March 3rd, 1760, the Flying Machine was started, and advertised to perform the journey, "if God permit," in three days, by John Hanforth Matthew Howe, Samuel Granville, and William Richardson. Fare inside, 2*l.* 5*s.*; outside, half-price.

siderable distances slung across horses' backs, were so dear as to be beyond the reach of the mass of the population. The distress caused by this frequent dearth of provisions was not effectually remedied until the canal navigation became completely opened up. Thus a great scarcity of food occurred in Manchester and the neighbourhood in 1757, which the common people attributed to the millers and corndealers; and unfortunately the notion was not confined to the poor who were starving, but was equally entertained by the well-to-do classes who had enough to eat. An epigram by Dr. Byrom, the town clergyman, written in 1737, on two millers (tenants of the School corn-mills), who, from their spare habits, had been nicknamed "Skin" and "Bone," was now revived, and tended to fan the popular fury. It ran thus:—

"Bone and Skin, two millers thin,
    Would starve the town, or near it;
But be it known to Skin and Bone,
    That Flesh and Blood can't bear it."

The popular hunger and excitement increasing, at length broke out in open outrage; and a riot took place in 1758, long after remembered in Manchester as the "Shude Hill fight," in which unhappily several lives were lost.

For the same reasons, the supply of coals was but scanty in winter; and though abundance of the article lay underground, within a few miles of Manchester, in nearly every direction, those few miles of transport, in the then state of the roads, were an almost insurmountable difficulty. The coals were sold at the pit mouth at so much the horse-load, weighing 280 lbs., and measuring two baskets, each thirty inches by twenty, and ten inches deep; that is, as much as an average horse could carry on its back.* The price of

---

* This "load" is still used as a measure of weight, though the practice of carrying all sorts of commodities on horses' backs, in which it originated, has long since ceased.

the coals at the pit mouth was 10*d*. the horse-load ; but by the time the article reached the door of the consumer in Manchester, the price was usually more than doubled, in consequence of the difficulty and cost of conveyance. The carriage alone amounted to about nine or ten shillings the ton.

There was as yet no connection of the navigation of the Mersey and Irwell with any of the collieries situated to the eastward of Manchester, by which a supply could reach the town in boats ; and although the Duke's collieries were only a comparatively short distance from the Irwell, the coals had to be carried on horses' backs or in carts from the pits to the river to be loaded, and after reaching Manchester they had again to be carried to the doors of the consumers,—so that there was little if any saving to be effected by that route. Besides, the minimum charge insisted on by the Mersey Navigation Company of 3*s*. 4*d*. a ton for even the shortest distance, proved an effectual barrier against any coal reaching Manchester by the river.

The same difficulty stood in the way of the transit of goods between Manchester and Liverpool. By road the charge was 40*s*. a ton, and by river 12*s*. a ton; that between Warrington and Manchester being 10*s*. a ton : besides, there was great risk of delay, loss, and damage by the way. Some idea of the tediousness of the river navigation may be formed from the fact, that the boats were dragged up and down stream exclusively by the labour of men, and that horses and mules were not employed for this purpose until after the Duke's canal had been made. It was, indeed, obvious that unless some means could be devised for facilitating and cheapening the cost of transport between the seaport and the manufacturing towns, there was little prospect of any considerable further development being effected in the industry of the district.

Such was the state of things when the Duke of Bridgewater turned his attention to the making of a water-road for the passage of his coal from Worsley to Manchester.

The Old Mersey Company would give him no facilities for sending his coals by their navigation, but levied the full charge of 3s. 4d. for every ton he might send to Manchester by river even in his own boats. He therefore perceived that to obtain a vend for his article, it was necessary he should make a way for himself; and it became obvious to him that if he could but form a canal between the two points, he would at once be enabled to secure a ready sale for all the coals that he could raise from his Worsley pits.

We have already stated that, as early as 1737, an Act had been obtained by the Duke's father, giving power to make the Worsley Brook navigable from the neighbourhood of the pits to the Irwell. But the enterprise, and its cost, appear to have been too formidable; so the powers of the Act were allowed to expire without anything being done to carry them out. The young Duke now determined to revive the Act in another form, and in the early part of 1759 he applied to Parliament for the requisite powers to enable him to cut a navigable canal from Worsley Mill eastward to Salford, and to carry the same westward to a point on the river Mersey, called Hollin Ferry. He introduced into the bill several important concessions to the inhabitants of Manchester. He bound himself not to exceed the freight of 2s. 6d. per ton on all coals brought from Worsley to Manchester, and not to sell the coal so brought from the mines to that town at more than 4d. per hundred, which was less than half the then average price. It was clear that, if such a canal could be made and the navigation opened as proposed, it would prove a great public boon to the inhabitants of Manchester. The bill was accordingly well supported, and it passed the legislature without opposition, receiving the Royal assent in March, 1759.

The Duke gave further indications of his promptitude and energy, in the steps which he adopted to have the works carried out without loss of time. He had no intention of allowing the powers of this Act to remain a dead

letter, as the former had done.   Accordingly, no sooner had
it passed than he set out for his seat at Worsley to take the
requisite measures for constructing the canal.   The Duke
was fortunate in having for his land-agent a very shrewd,
practical, and enterprising person, in John Gilbert, whom
he consulted on all occasions of difficulty.

Mr. Gilbert was the brother of Thomas Gilbert, the
originator of the Gilbert Unions, then agent to the Duke's
brother-in-law, Lord Gower.   That nobleman had for some
time been promoting the survey of a canal to unite the
Mersey and the Trent, on which Brindley had been employed, and thus became known to Gilbert as well as to
his brother.   We find from an entry in Brindley's pocket-
book that the millwright had sundry interviews with
Thomas Gilbert on matters of business previous to the
passing of the first Bridgewater Canal Bill, though there
is no evidence that he was employed in making the survey.
Indeed, it is questionable whether any survey was made of
the first scheme.   Engineering projects were then submitted
to Parliamentary Committees in a very rough state.   Levels
were guessed at rather than surveyed and calculated; and
merely general powers were taken enabling such property
to be purchased as might by possibility be required for the
execution of the works.   In the case of the Bridgewater
Canal, the prices of land and compensation for damage
were directed to be assessed by a local committee appointed
by the Act for the purpose.

When the Duke proceeded to consider with Gilbert the
best mode of carrying out the proposed canal, it appeared
clear to them that the plan originally contemplated
was faulty in many respects, and that an application must
be made to Parliament for further powers.   By the
original Act it was intended to descend from the level
of the coal-mines at Worsley by a series of locks into
the river Irwell.   This, it was found, would necessarily
involve a heavy cost both in the construction and working
of the canal, as well as considerable delay in the conduct

of the traffic, which it was most desirable to avoid. Neither the Duke nor Gilbert had any practical knowledge of engineering; nor, indeed, were there many men in the

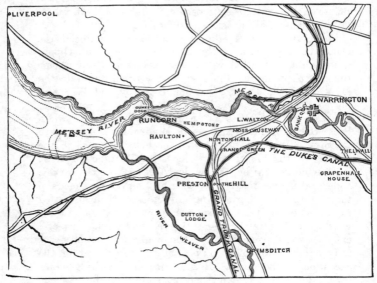

Map of the Duke's Canal.
[Western Part.]

country at that time who knew much of the subject; for it must be remembered that this canal of the Duke's was the very first project in England for cutting a navigable trench through the dry land, and carrying merchandise in it across the country, independent of the course of the existing streams.

It was in this emergency that Gilbert advised the Duke to call to his aid James Brindley, whose fertility of resources and skill in overcoming mechanical difficulties had long been the theme of general admiration in his own district. Doubtless the Duke was as much impressed by the native vigour and originality of the unlettered genius introduced to him by his agent, as were all with whom he was brought in contact. Certain it is that the

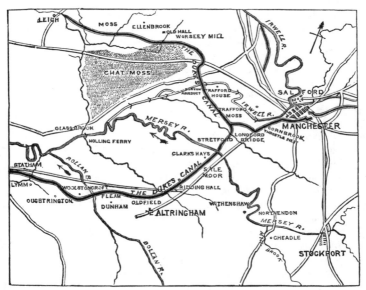

Map of the Duke's Canal.
[Eastern Part.]

Duke showed his confidence in Brindley by entrusting him with the conduct of the proposed work; and, as the first step, he was desired to go over the ground at once, and give his opinion as to the best plan to be adopted for carrying it out with despatch.

Brindley, accordingly, after making what he termed an " ochilor [ocular] servey or a ricconitoring," speedily formed his conclusion, and came back to the Duke with his advice. It was that, instead of carrying the canal down into the Irwell by a flight of locks, and so up again on the other side to the proposed level, it should be carried

right over the river, and constructed upon one uniform
level throughout. But this, it was clear, would involve
a series of formidable works, the like of which had not
before been attempted in England. In the first place,
the low ground on the north side of the Irwell would
have to be filled up by a massive embankment, and to
be united with the land on the other bank by means of
a large aqueduct of stone. Would it be practicable or
possible to execute works of such magnitude? Brindley
expressed so strong and decided an opinion of their practicability, that the Duke was won over to his views, and
determined again to go to Parliament for the requisite
powers to enable him to carry out the design.

Many were the deliberations which took place about
this time between the Duke, Gilbert, and Brindley, in
the Old Hall at Worsley, where the Duke had now taken
up his abode. We find from Brindley's pocket-book memoranda, that in the month of July, 1759, he had taken up
his temporary quarters at the Old Hall; and from time
to time, in the course of the same year, while the details
of the plan were being prepared with a view to the
intended application to Parliament, he occasionally stayed
with the Duke for weeks together. He made a detailed
survey of the new line, and at the same time, in order to
facilitate the completion of the undertaking when the
new powers had been obtained, he proceeded with the
construction of the sough or level at Worsley Mill, and
such other portions of the work as could be executed
under the original powers.

During the same period Brindley travelled backwards
and forwards a great deal, on matters connected with
his various business in the Pottery district. We find,
from his private record, that he was occupied at intervals
in carrying forward his survey of the proposed canal
through Staffordshire, visiting with this object the neighbourhood of Newcastle-under-Lyne, Lichfield, and Tam-
worth. He also continued to give his attention to mills,

water-wheels, cranes, and fire-engines, which he had
erected, or which required repairs, in various parts of
the same district. In short, he seems at this period to have
been fully employed as a millwright; and although, as
we have seen, the remuneration which he received for
his skill was comparatively small, being a man of frugal
habits he had saved a little money; for about this time
we find him able to raise a sum of 543*l.* 6*s.* 8*d.*, being
his fourth share of the purchase-money of the Turnhurst
estate, situated near Golden Hill, in the county of Stafford.

The principal part of this sum was no doubt borrowed,
as appears by his own memoranda, from his friend Mr.
Launcelot, of Leek; but the circumstance proves that,
amongst his townsmen and neighbours, who knew him
best, he stood in good credit and repute. His other
partners in the purchase were Mr. Thomas Gilbert (Earl
Gower's agent), Mr. Henshall (afterwards his brother-in-
law), and his brother John Brindley. The estate was
understood to be full of minerals, the knowledge of which
had most probably been obtained by Brindley in the
course of his surveying of the proposed Staffordshire canal;
and we shall afterwards find that he turned the purchase
to good account.

At length the new plans of the canal from Worsley to
Manchester were completed and ready for deposit; and
on the 23rd of January, after a visit to the Duke and
Gilbert at the Hall, we find the entry in Brindley's
pocket-book of "Sot out for London." On the occasion
of his visits to London, Brindley adopted the then most
convenient method of travelling on horseback, the journey
usually occupying five days. We find him varying his
route according to the state of the weather and of the
roads. In summer he was accustomed to go by Coventry,
but in winter he made for the Great North Road by Northampton, which was usually in better condition for winter
travelling.

The second Act passed like the first, without opposition,

early in the session of 1760. It enabled the Duke to carry his proposed canal *over* the river Irwell, near Barton Bridge, some five miles westward of Manchester, by means of a series of arches, and to vary its course accordingly; whilst it further authorised him to extend a short branch to Longford Bridge, near Stretford,—that to Hollin Ferry, authorised by the original Act, being abandoned. In the mean time the works near Worsley had been actively pushed forward, and considerable progress had been made by the time the additional powers had been obtained. That part of the canal which lay between Worsley Mill and the public highway leading from Manchester to Warrington had been cut; the sough or level between Worsley Mill and Middlewood, for the purpose of supplying water to the canal, was considerably advanced; and operations had also been begun in the neighbourhood of Salford and on the south of the river Irwell.

The most difficult part of the undertaking, however, was that authorised by the new Act; and the Duke looked forward to its execution with the greatest possible anxiety. Although aqueducts of a far more formidable description had been executed abroad, nothing of the kind had until then been projected in this country; and many regarded the plan of Brindley as altogether wild and impracticable. The proposal to confine and carry a body of water within a water-tight trunk of earth upon the top of an embankment across the low grounds on either side of the Irwell, was considered foolish and impossible enough; but to propose to carry ships upon a lofty bridge, over the heads of other ships navigating the Irwell which flowed underneath, was laughed at as the dream of a madman. Brindley, by leaving the beaten path, thus found himself exposed to the usual penalties which befall originality and genius.

The Duke was expostulated with by his friends, and strongly advised not to throw away his money upon so desperate an undertaking. Who ever heard of so large a body of water being carried over another in the man ner proposed? Brindley was himself appealed to; but he could only repeat his conviction as to the entire practicability of his design. At length, by his own desire and to allay the Duke's apprehensions, another engineer was called in and consulted as to the scheme. To Brindley's surprise and dismay, the person consulted concurred in the view so strongly expressed by the public. He characterised the plan of the Barton aqueduct and embankment as instinct with recklessness and folly; and after expressing his unqualified opinion as to the impracticability of executing the design, he concluded his report to the Duke thus: "I have often heard of castles in the air; but never before saw where any of them were to be erected." *

It is to the credit of his Grace that, notwithstanding these strong adverse opinions, he continued to give his confidence to the engineer whom he had selected to carry out the work. Brindley's common-sense explanations, though they might not remove all his doubts, nevertheless determined the Duke to give him the full opportunity of carrying out his design ; and he was accordingly authorised to proceed with the erection of his " castle in the air." Its progress was watched with great interest, and people flocked from all parts to see it.

The Barton aqueduct is about two hundred yards in length and twelve yards wide, the centre part being sustained by a bridge of three semicircular arches, the middle one being of sixty-three feet span. It carries the canal over the Irwell at a height of thirty-nine feet above the river—this head-room being sufficient to enable the largest barges to pass underneath without lowering their masts. The bridge is entirely of stone blocks, those on

---

* We have heard the name of Smeaton mentioned as that of the engineer consulted on the occasion, but are unable to speak with certainty on the point. Excepting Smeaton, however, there was then no other engineer in the country of recognised eminence in the profession.

the faces being dressed on the front, beds, and joints, and cramped with iron. The canal, in passing over the arches, is confined within a puddled * channel to prevent leakage, and is in as good a state now as on the day on which it was completed. Although the Barton aqueduct has since been thrown into the shade by the vastly greater works of modern engineers, it was unquestionably a very bold and ingenious enterprise, if we take into account the time at which it was erected. Humble though it now appears, it was the parent of the magnificent aqueducts of Rennie and Telford, and of the viaducts of Stephenson and Brunel, which rival the greatest works of any age or country.

The embankments formed across the low grounds on either side of the Barton viaduct were also considered very formidable works at that day. A contemporary writer speaks of the embankment across Stretford Meadows as an amazing bank of earth 900 yards long, 112 feet in breadth across the base, 24 feet at top, and 17 feet high. The greatest difficulty anticipated, was the holding of so large a body of water within a hollow channel formed of soft materials. It was supposed at first that the water would soak through the bank, which its weight would soon burst, and wash away all before it. But Brindley, in the course of his experience, had learnt something of the powers of clay-puddle to resist the passage of water. He had already succeeded in stopping the breaches of rivers flowing through low grounds by this means; and the thorough

---

* The process of puddling is of considerable importance in canal engineering. Puddle is formed by a mixture of well-tempered clay and sand reduced to a semi-fluid state, and rendered impervious to water by manual labour, as by working and chopping it about with spades. It is usually applied in three or more strata to a depth or thickness of about three feet: and care is taken at each operation so to work the new layer of puddling stuff as to unite it with the stratum immediately beneath. Over the top course a layer of common soil is usually laid. It is only by the careful employment of puddling that the filtration of the water of canals into the neighbouring lower lands through which they pass can be effectually prevented.

Barton Aqueduct.

[By Percival Skelton, after his original Drawing.]

manner in which he finished the bed of this canal, and made it impervious to water, may be cited as a notable illustration of the engineer's practical skill, taking into account the early period at which this work was executed.

Not the least difficult part of the undertaking was the formation of the canal across Trafford Moss, where the weight of the embankment pressed down and "blew up" the soft oozy stuff on each side; but the difficulty was again overcome by the engineer's specific of clay-puddle, which proved completely successful. Indeed, the execution of these embankments by Brindley was regarded at the time as something quite as extraordinary in their way as the erection of the Barton aqueduct itself.

The rest of the canal between Longford and Manchester, being mostly on sidelong ground, was cut down on the upper side and embanked up on the other by means of the excavated earth. This was comparatively easy work; but a matter of greater difficulty was to accommodate the streams which flowed across the course of the canal. This was, however, provided for in a highly ingenious manner. For instance, a stream called Cornbrook was found too high to pass under the canal at its natural level. Accordingly, Brindley contrived a weir, over which the stream fell into a large basin, from whence it flowed into a smaller one open at the bottom. From this point a culvert, constructed under the bed of the canal, carried the waters across to a well situated on its further side, where the waters rising up to their natural level, again flowed away in their proper channel. A similar expedient was adopted at the Manchester terminus of the canal, at the point at which it joined the waters of the Medlock.

It was a principle of Brindley's never to permit the waters of any river or brook to intermix with those of the canal except for the purpose of supply; as it was clear that in a time of flood such intermingling would be a source of great danger to the navigation. In order, therefore, to provide for the free passage of the Medlock without causing a rush into the canal, a weir was contrived 366 yards in circumference, over which its waters flowed into a lower level, and from thence into a well several yards in depth, down which the whole river fell. It was received at the bottom in a subterranean passage, by which it passed into the river Irwell, near at hand. The weir was very ingeniously contrived, though it was afterwards found necessary to make considerable alterations and improvements in it, as experience suggested, in order effectually to accommodate the flood-waters of the Medlock. Arthur Young, when visiting the canal, shortly after it was opened up to Manchester, says, "The whole plan of these works shows a capacity and extent of mind which foresees difficulties, and invents remedies in anticipation of possible evils. The connection and dependence of the parts upon each other are happily imagined; and all are exerted in concert, to command by every means the wished-for success." *

Brindley's labours, however, were not confined to the construction of the canal, but his attention seems to have been equally directed to the contrivance of the whole of the arrangements and machinery by which it was worked. The open navigation between Worsley Mill and Manchester was 10¼ miles in length. A large basin was excavated at the former place, of sufficient capacity to contain a great many boats, and to serve as a head for the navigation.

It is at Worsley Basin that the canal enters the bottom of the hill by a subterranean channel which extends for a great distance,—connecting the different workings of the mine,—so that the coals can be readily transported in boats to their place of sale. A representation of the basin is given in the annexed cut. It lies at the base of

---

* 'Six Months' Tour through the North of England.' vol. iii., p. 258. Ed. 1770.

a cliff of sandstone, some hundred feet in height, over-hung by luxuriant foliage, beyond which is seen the graceful spire of Worsley church. In contrast to this scenic beauty above, lies the almost stagnant pool beneath. The barges* laden with coal emerge from the mine through the two low, semi-circular arches opening at the base of the rock, such being the entrances to the underground workings. The smaller aperture is the mouth of a canal of only half a mile in length, serving to prevent the obstruction which would be caused by the entrance and egress of so many barges through a single passage. The other archway is the entrance of a wider channel, extending nearly six miles in the direction of Bolton, from which various other canals diverge in different directions.

In Brindley's time, this subterranean canal, hewn out of the rock, was only about a mile in length, but it now extends to nearly forty miles in all directions underground. Where the tunnel passed through earth or coal, the arching was of brickwork; but where it passed through rock, it was simply hewn out. This tunnel acts not only as a drain and water-feeder for the canal itself, but as a means of carrying the facilities of the navigation through the very heart of the collieries; and it will readily be seen of how great a value it must have proved in the economical working of the navigation, as well as of the mines, so far as the traffic in coals was concerned.

At every point Brindley's originality and skill were at work. He invented the cranes for the purpose of more readily loading the boats with the boxes filled with the Duke's "black diamonds." He also contrived and laid down within the mines a system of underground railways, all leading from the face of the coal, where the miners

---

* The barges are narrow and long, each conveying about ten tons of coal. They are drawn along the tunnels by means of staples fastened to the sides. When they are empty, and consequently higher in the water, they are so near the roof that the bargemen, lying on their backs, can propel them with their feet.

Worsley Basin.

worked, to the wells which he had made at different points
in the tunnels, through which the coals were shot into the
boats waiting below to receive them. At Manchester,
where they were unloaded for sale, the contrivances which
he employed were equally ingenious. It was at first
intended that the canal should terminate at the foot of
Castle Hill, up which the coals were dragged by their
purchasers from the boats in wheelbarrows or carts. But
the toil of dragging the loads up the hill was found very
great; and, to remedy the inconvenience, Brindley contrived to extend the canal for some way into the hill,
opening a shaft from the surface of the ground down to the
level of the water. The barges having made their way to
the foot of this shaft, the boxes of coal were hoisted to the
surface by a crane, worked by a box water-wheel of 30 feet
diameter and 4 feet 4 inches wide, driven by the waterfall
of the river Medlock. In this contrivance Brindley was
only adopting a modification of the losing and gaining
bucket, moved on a vertical pillar, which he had before
successfully employed in drawing water out of coal-mines.
By these means the coals were rapidly raised to the higher
ground, where they were sold and distributed, greatly to
the convenience of those who came to purchase them.

Brindley's practical ability was equally displayed in
planning and building a viaduct or in fitting up a crane
—in carrying out an embankment or in contriving a coal-
barge. The range and fertility of his constructive genius
were extraordinary. For the Duke, he invented water-
weights at Rough Close, riddles to wash coal for the forges,
raising dams, and numerous other contrivances of well-
adapted mechanism. At Worsley he erected a steam-
engine for draining those parts of the mine which were
beneath the level of the canal, and consequently could not
be drained into it; and he is said to have erected, at a cost
of only 150l., an engine which until that time no one had
known how to construct for less than 500l. At the mouth
of one of the mines he erected a water-bellows for the pur-

pose of forcing fresh air into the interior, and thus venti-
lating the workings.*   At the entrance of the underground
canal he designed and built a mill of a new construction, driven by an over-shot wheel twenty-four feet in
diameter, which worked three pair of stones for grinding
corn, besides a dressing or boulting mill, and a machine for
sifting sand and mixing mortar.

Brindley's quickness of observation and readiness in
turning circumstances to advantage were equally displayed
in the mode by which he contrived to obtain an ample
supply of lime for building purposes during the progress of
the works.   We give the account as related by Arthur
Young :—" In carrying on the navigation," he observes, " a
vast quantity of masonry was necessary for building
aqueducts, bridges, warehouses, wharves, &c., and the
want of lime was felt severely.   The search that was made
for matters that would burn into lime was for a long time
fruitless.   At last Mr. Brindley met with a substance of a
chalky kind, which, like the rest, he tried; but found
(though it was of a limestone nature—lime-marl, which
was found along the sides of the canal, about a foot below
the surface) that, for want of adhesion in the parts, it
would not make lime.   This most inventive genius happily
fell upon an expedient to remedy this misfortune.   He
thought of tempering this earth in the nature of brick-
earth, casting it in moulds like bricks, and then burning
it; and the success was answerable to his wishes.   In that

---

* A writer in the 'St. James's
Chronicle,' under date the 30th of
September, 1763, gives the following account of this apparatus, long
since removed :—" At the mouth of
the cavern is erected a water-
bellows, being the body of a tree,
forming a hollow cylinder, standing
upright.   Upon this a wooden basin
is fixed, in the form of a funnel,
which receives a current of water
from the higher ground.   This
water falls into the cylinder, and
issues out at the bottom of it, but
at the same time carries a quantity
of air with it, which is received
into the pipes and forced to the
innermost recesses of the coalpits,
where it issues out as if from a
pair of bellows, and rarefies the
body of thick air, which would
otherwise prevent the workmen
from subsisting on the spot where
the coals are dug."

state it burnt readily into excellent lime; and this acquisition was one of the most important that could have been made. I have heard it asserted more than once that this stroke was better than twenty thousand pounds in the Duke's pocket; but, like most common assertions of the same kind, it is probably an exaggeration. However, whether the discovery was worth five, ten, or twenty thousand, it certainly was of noble use, and forwarded all the works in an extraordinary manner." *

It has been stated that Brindley's nervous excitement was so great on the occasion of the letting of the water into the canal, that he took to his bed at the Wheatsheaf, in Stretford, and lay there until all cause for apprehension was over. The tension on his brain must have been great, with so tremendous a load of work and anxiety upon him; but that he "ran away," † as some of his detractors have

---

* 'Six Months' Tour,' vol. iii., p. 270–1. Mr. Hughes, C.E., says of this discovery : "The lime thus made would appear to be the first cement of which we have any knowledge in this country ; since the calcareous marl here spoken of would probably produce, when burnt, a lime of strong hydraulic properties."

† This story was first set on foot, we believe, by the Earl of Bridgewater, in his singularly incoherent publication entitled, 'A Letter to the Parisians and the French Nation upon Inland Navigation, containing a defence of the public character of His Grace Francis Egerton, late Duke of Bridgewater. By the Hon. Francis Henry Egerton.' The first part of this curious book (published at Paris) was dated "Hôtel Egerton, Paris, 21st Dec., 1818;" the second part was published two years later; and a third part, consisting entirely of a note about Hebrew interpretations, was published subsequently. He had in the mean time become Earl of Bridgewater, in October, 1823, having formerly been prebendary of Durham and rector of Whitchurch in Shropshire. The late Earl of Ellesmere, in his 'Essays on History, Biography,' &c., says of this nobleman that "he died at Paris in the odour of eccentricity." But this is a mild description of his lordship, who had at least a dozen distinct crazes—about canals, the Jews, punctuation, the wonderful merits of the Egertons, the proper translation of Hebrew, the ancient languages generally, but more especially about prophecy and poodle-dogs. When he drove along the Boulevards in Paris, nothing could be seen of his lordship for poodle-dogs looking out of the carriage-windows. The poodles sat at table with him at dinner, each being waited on by a special valet. The most creditable thing the Earl did was to leave the sum of 12,000*l*

alleged, is at variance with the whole character and history of the man.

The Duke's canal, when finished, was for a long time regarded as the wonder of the neighbourhood. Strangers flocked from a distance to see Brindley's "castle in the air;" and contemporary writers spoke in glowing terms of the surprise with which they saw several barges of great burthen drawn by a single mule or horse along "a river hung in the air," over another river flowing underneath, by the side of which some ten or twelve men might be seen slowly hauling a single barge against the stream. A lady who writes a description of the work in 1765, speaks of it as "perhaps the greatest artificial curiosity in the world;" and she states that "crowds of people, including those of the first fashion, resort to it daily."

The chief importance of the work, however, consisted in its valuable uses. Manchester was now regularly and cheaply supplied with coals. The average price was at once reduced by one-half—from 7d. the cwt. to 3½d. (six score being given to the cwt.)—and the supply was regular instead of intermitting, as it had formerly been. But the full advantages of this improved supply of coals were not experienced until many years after the opening of the canal, when the invention of the steam-engine, and its extensive employment as a motive power in all manufacturing operations, rendered a cheap and abundant supply of fuel of vital importance to the growth and prosperity of Manchester and its neighbourhood.

----

to the British Museum, and 8000l. to meritorious literary men for writing the well-known 'Bridge-water Treatises.' He died in February, 1829.

CHAPTER IV.

Extension of the Duke's Canal to the Mersey.

The Canal had scarcely been opened to Manchester when we find Brindley occupied, at the instance of the Duke, in surveying the country between Stretford and the river Mersey, with the view of carrying out a canal in that direction for the accommodation of the growing trade between Liverpool and Manchester. The first boat-load of coals sailed over the Barton viaduct to Manchester on the 17th of July, 1761; on the 7th of September following we find Brindley at Liverpool,* "rocconitoring;" and, by the end of the month, he was busily engaged levelling for a proposed canal to join the Mersey at Hemp-stones, about eight miles below Warrington Bridge, from whence there was a natural tideway to Liverpool, about fifteen miles distant.

The project in question was a very important one on public grounds. We have seen how the community of Manchester had been hampered by defective road and water communications, which seriously affected its supplies of food and fuel, and, at the same time, by retarding its trade, hindered to a considerable extent the regular em-

---

* It would almost seem as if the extension of the canal to the Mersey had formed part of the Duke's original plan; for Brindley was engaged in making a survey from Longford to Dunham in the autumn of the preceding year, as appears from the following account, preserved at the Bridgewater Canal Office at Manchester, of his expenses in making the survey :—

"Expenses in Surveying from Longford Bridge to Dunham.

Octr 21st 1760.		
Spent at Stretford	0	5
At Altringham all Night	6	0
Gave the Men to Drink that assisted .. .. }	1	0
22nd		
More at Altringham	2	6
	10	0

P^d Mr. Brinley this."

ployment of its population. The Duke of Bridgewater,
by constructing his canal, had opened up an abundant
supply of coal, but the transport of the raw materials of
manufacture was still as much impeded as before. Liverpool was the natural port of Manchester, from which it
drew its supplies of cotton, wool, silk, and other produce,
and to which it returned them for export when worked up
into manufactured articles.

There were two existing modes by which the communication was kept up between the two places: one was by
the ordinary roads, and the other by the rivers Mersey and
Irwell. From a statement published in December, 1761, it
appears that the weight of goods then carried by land
from Manchester to Liverpool was " upwards of forty tons
per week," or about two thousand tons a year. This
quantity, insignificant though it must appear when compared with the enormous traffic now passing between
the two towns, was then thought very large, as no doubt
it was when the limited trade of the country is taken into
account. But the cost of transport was the important
feature; it was not less than two pounds sterling per ton
—this heavy charge being almost entirely attributable to
the execrable state of the roads. It was scarcely possible
to drive waggons along the ruts and through the sloughs
which lay between the two places at certain seasons of the
year, and even pack-horses had considerable difficulty in
making the journey.

The other route between the towns was by the navigation of the rivers Mersey and Irwell. The raw materials
used in manufacture were principally transported from
Liverpool to Manchester by this route, at a cost of
about twelve shillings per ton; the carriage of timber and
such like articles costing not less than twenty per cent. on
their value at Liverpool. But the navigation was also
very tedious and difficult. The boats could only pass up
to the first lock at the Liverpool end with the assistance
of a spring tide; and further up the river there were

numerous fords and shallows which the boats could only pass in great freshes, or, in dry seasons, by drawing extraordinary quantities of water from the locks above. Then, in winter, the navigation was apt to be impeded by floods, and occasionally it was stopped altogether. In short, the growing wants of the population demanded an improved means of transit between the two towns, which the Duke of Bridgewater now determined to supply.

The growth of Liverpool as a seaport has been comparatively recent. At a time when Bristol and Hull possessed thriving harbours, resorted to by foreign ships, Liverpool was little better than a fishing village, its only distinction being that it was a convenient place for setting sail to Ireland. In the war between France and England which broke out in 1347, when Edward the Third summoned the various ports in the kingdom to make contributions towards the naval power according to their means, London was required to provide 25 ships and 662 men; Bristol 22 ships and 608 men; Hull, 16 ships and 466 men; whilst Liverpool was only asked to find 1 bark and 6 men! In Queen Elizabeth's time, the burgesses presented a petition to Her Majesty, praying her to remit a subsidy which had been imposed upon it and other seaport towns, in which they styled their native place " Her Majesty's poor decayed town of Liverpool." Chester was then of considerably greater importance as a port. In 1634-5, when Charles I. made his unconstitutional levy of ship-money throughout England, Liverpool was let off with a contribution of 15l., whilst Chester paid 100l., and Bristol not less than 1000l.

The channel of the Dee, however, becoming silted up, the trade of Chester decayed, and that of Liverpool rose upon its ruins. In 1699 the excavation of the Old Dock was begun; but it was used only as a tidal harbour (being merely an enclosed space with a small pier) until the year 1709, when an Act was obtained enabling its conversion into a wet dock; since which time a series of docks have been constructed, extending for about five miles along the north shore of the Mersey, which are among the greatest works of modern times, and afford an almost unequalled amount of shipping accommodation.

Liverpool in 1650.

From that time forward the progress of the port of Liverpool has kept steady pace with the trade and wealth of the country behind it, and especially with the manufacturing activity and energy of the town of Manchester. Its situation at the mouth of a deep and navigable river, its convenient proximity to districts abounding in coal and iron and inhabited by an industrious and hardy population, were unquestionably great advantages. But these of themselves would have been insufficient to account for the extraordinary progress made by Liverpool during the last century, without the opening up of the great system of canals, which brought not only the towns of Yorkshire, Cheshire, and Lancashire into immediate connection with that seaport, but also the manufacturing districts of Staffordshire, Warwickshire, and the other central counties of England situated at the confluence of the various navigations.* Liverpool thus became the great focus of import and export for the northern and western districts. The raw materials of commerce were poured into it from Ireland, America, and the Indies. From thence they were distributed along the canals amongst the various seats of manufacturing industry, and a large proportion was readily returned by the same route to the same port, in a manufactured state, for shipment to all parts of the world.

At the time of which we speak, however, it will be observed that the communication between Liverpool and Manchester was very imperfect. It was not only difficult to convey goods between the two places, but it was also difficult to convey persons. In fine weather, those who required to travel the thirty miles which separated them, could ride or walk, resting at Warrington for the night. But in winter the roads, like most of the other country roads at the time, were simply impassable. Although an Act had been passed as early as the year 1726 for repairing and enlarging the road from Liverpool to Prescot, coaches could not come nearer to the town than Warrington in 1750, the road being impracticable for such vehicles even in summer.†

A stage-coach was not started between Liverpool and Manchester until the year 1767, performing the journey only three times a week. It required six and sometimes eight horses to draw the lumbering vehicle and its load

---

* Progress of Liverpool.

Years.	Vessels entered.	Tonnage.	Duties Paid.
1701	102	8,619	..
1760	1,245	..	£2,330
1800	4,746	450,060	23,379
1858	21,352	4,441,943	347,889

† Mr. Baines says : " Carriages were then very rare, and it is mentioned as a singular fact that at the period in question (1750) there was but one *gentleman's* carriage in the town of Liverpool, and that carriage was kept by a *lady* of the name of Clayton."—'History of Lancashire,' vol. iv.. p. 90.

along the ruts and through the sloughs,—the whole day
being occupied in making the journey. The coach was
accustomed to start early in the morning from Liverpool ;
it breakfasted at Prescot, dined at Warrington, and arrived
at Manchester usually in time for supper. On one occasion,
at Warrington, the coachman intimated his wish to proceed,
when the company requested him to take another pint, as
they had not finished their wine, asking him at the same
time if he was in a hurry ? "Oh," replied the driver, "I'm
not partic'lar to an hour or so !" As late as 1775, no mail-
coach ran between Liverpool and any other town, the bags
being conveyed to and from it on horseback ; and one
letter-carrier was found sufficient for the wants of the
place. A heavy stage then ran, or rather crawled, between
Liverpool and London, making only four journeys a week
in the winter time. It started from the Golden Talbot, in
Water-street, and was three days on the road. It went by
Middlewich, where one of its proprietors kept the White
Bear inn ; and during the Knutsford race-week the coach
was sent all the way round by that place, in order to bring
customers to the Bear.

We have said that Brindley was engaged upon the
preliminary survey of a canal to connect Manchester with
the Mersey, immediately after the original Worsley line
had been opened, and before its paying qualities had
been ascertained. But the Duke, having once made
up his mind as to the expediency of carrying out this
larger project, never halted nor looked back, but made
arrangements for prosecuting a bill for the purpose of
enabling the canal to be made in the very next session
of Parliament.

We find that Brindley's first visit to Liverpool and the
intervening district on the business of the survey was
made early in September, 1761. During the remainder
of the month he was principally occupied in Staffordshire,
looking after the working of his fire-engine at Fenton
Vivian, carrying out improvements in the silk-manufactory

at Congleton, and inspecting various mills at Newcastle-
under-Lyne and the neighbourhood. His only idle day
during that month seems to have been the 22nd, which was a
holiday, for he makes the entry in his book of " crounation
of Georg and Sharlot," the new King and Queen of England. By the 25th we find him again with the Duke at
Worsley, and on the 30th he makes the entry, " set out at
Dunham to Level for Liverpool." The work then went
on continuously until the survey was completed; and on
the 19th of November he set out for London, with 7*l.* 18*s.*
in his pocket.

In the course of his numerous journeys, we find Brindley carefully noting down the various items of his expenses,
which were curiously small. Although he was four or
five days on the road to London, and stayed eight days
there, his total expenses, both going and returning,
amounted to only 4*l.* 8*s.*: it is most probable, however, that
he lived at the Duke's house whilst in town. On the
1st of December we find him, on his return journey to
Worsley, resting the first night at a place called Brick-
hill; the next at Coventry, where he makes the entry,
" Moy mar had a bad fall in the frasst;" the third at
Sandon; the fourth at Congleton; and the fifth at Worsley.
He had still some inquiries to make as to the depth of
water and the conditions of the tide at Hempstones; and
for three days he seems to have been occupied in traffic-
taking, with a view to the evidence to be given before
Parliament; for on the 10th of December we find him
at Stretford, " to count the caridgos," and on the 12th he
is at Manchester for the same purpose, " counting the
loded caridgos and horses."

The following bill refers to some of the work done by
him at this time, and is a curious specimen of an engineer's
travelling charges in those days—the engineer himself
being at the same time paid at the rate of 3*s.* 6*d.* a day :—

*Expenses for His Grace the Duk of Bridgwator to pay for traveling
Chareges by James Brindley.*

18 Novem—1761.

18	No masuring a Cros from Dunham to Warbuton Mercey and Thalwall, 3s - 11d Dunham for 2 diners 1s - 3d for the man 1s - 0d at Thalwall 1s - 2d all Night Warington	0	7	4
19	Novem Sat out from Chester for London & at Worsley Septm 5 Retorned back going to London and at London & hors back to Worsley Charged Hors & my salf	4	8	0
9	december Coming back from Ham Stone Charges at Wilderspool all Night	0	8	0
	at Warington to meet Mr Ashley dining	0	4	2
10	to ataind the Turn pike Rode 2s - 6d & againe on te 12 De Rode 3s - 6d	0	6	0
21	Decm to inspect te flux and Reflux at Ham Stone 2 dayes Charges	0	6	6
26	Decr 1761. Recd the Contents of the above Bill by the Hands of John Gilbert. James Brindley.. .. .. .. £6	00	0

In the early part of the month of January, 1762, we
find Brindley busy measuring soughs, gauging the tides
at Hempstones, and examining and altering the Duke's
paper-mills and iron slitting-mills at Worsley; and on
the 7th we find this entry: " to masuor the Duks pools I
and Smeaton." On the following day he makes "an
ochilor survey from Saldnoor [Sale Moor] to Stockport,"
with a view to a branch canal being carried in that
direction. On the 14th, he sets out from Congleton, by
way of Ashbourne, Northampton, and Dunstable, arriving
in London on the fifth day.

Immediately on his arrival in town we find him proceeding to rig himself out in a new suit of clothes. His
means were small, his habits thrifty, and his wardrobe
scanty; but as he was about to appear in an important
character, as the principal engineering witness before
a Parliamentary Committee in support of the Duke's
bill, he felt it necessary to incur an extra expenditure on
dress for the occasion. Accordingly, on the morning of
the 18th we find him expending a guinea—an entire week's
pay—in the purchase of a pair of new breeches: two

guineas on a coat and waistcoat of broadcloth, and six shillings for a pair of new shoes. The subjoined is a fac-simile of the entry in his pocketbook.

Fac-simile of Brindley's Hand-writing.

It will be observed that an expenditure is here entered of nine shillings for going to "the play." It would appear that his friend Gilbert, who was in London with him on the canal business, prevailed on Brindley to go with

him to the theatre to see Garrick in the play of
'Richard III.,' and he went. He had never been to an
entertainment of the kind before; but the excitement
which it caused him was so great, and it so completely
disturbed his ideas, that he was unfitted for business for
several days after. He then declared that no consideration should tempt him to go a second time, and he held
to his resolution. This was his first and only visit to the
play. The following week he enters in his memorandum-
book concerning himself "ill in bed," and the first Sunday
after his recovery we find him attending service at "Sant
Mary's Church." The service did not make him ill, as
the play had done, and on the following day he attended
the House of Commons on the subject of the Duke's bill.

The proposed canal from Manchester to the Mersey
at Hempstones stirred up an opposition which none of the
Duke's previous bills had encountered. Its chief opponents
were the proprietors of the Mersey and Irwell Navigation,
who saw their monopoly assailed by the measure; and,
unable though they had been satisfactorily to conduct
the then traffic between Liverpool and Manchester, they
were unwilling to allow of any additional water service
being provided between the two towns. Having already
had sufficient evidence of the Duke's energy and enterprise, from what he had been able to effect in so short
a time in forming the canal between Worsley and Manchester, the Navigation Company were not without reason
alarmed at his present project.

At first they tried to buy him off by concessions. They
offered to reduce the rate of 3s. 4d. per ton of coals, timber,
&c., conveyed upon the Irwell between Barton and Manchester, to 6d. if he would join their Navigation at Barton
and abandon the part of his canal between that point
and Manchester; but he would not now be diverted from
his plan, which he resolved to carry into execution if
possible. Again they tried to conciliate his Grace by
offering him certain exclusive advantages in the use of

their Navigation. But it was again too late; and the Duke having a clear idea of the importance of his project, and being assured by his engineer of its practicability and the great commercial value of the undertaking, determined to proceed with the measure. It offered to the public the advantages of a shorter line of navigation, not liable to be interrupted by floods on the one hand or droughts on the other, and, at the same time, a much lower rate of freight, the maximum charge proposed in the bill being 6s. a ton against 12s., the rate charged by the Mersey and Irwell Navigation between Liverpool and Manchester.

The opposition to the bill was led by Lord Strange, son of the Earl of Derby, one of the members for the county of Lancaster, who took the part of the "Old Navigators," as they were called, in resisting the bill. The question seems also to have been treated as a political one; and, the Duke and his friends being Whigs, Lord Strange mustered the Tory party strongly against him. Hence we find this entry occurring in Brindley's note-book, under date the 16th of February: "The Toores [Tories] mad had [made head] agane ye Duk."

The principal objections offered to the proposed canal were, that the landowners would suffer by it from having their lands cut through and covered with water, by which a great number of acres would be for ever lost to the public; that there was no necessity whatever for the canal, the Mersey and Irwell Navigation being sufficient to carry more goods than the trade then existing required; that the new navigation would run almost parallel with the old one, and offered no advantage to the public which the existing river navigation did not supply; that the canal would drain away the waters which supplied the rivers, and be very prejudicial to them, if not totally destructive, in dry seasons; that the proprietors of the old navigation had invested their money on the faith of protection by Parliament, and to permit the new canal to be established would be a gross interference with their vested rights; and so on.

To these objections there were very sufficient answers. The bill provided for full compensation being made to the owners of lands through which the canal passed, and, in addition, it was provided that all sorts of manure should be carried for them without charge. It was also shown that the Duke's canal could not abstract water from either the Mersey or the Irwell, as the level of both rivers was considerably below that of the intended canal, which would be supplied almost entirely from the drainage of his own coal-mines at Worsley; and with respect to the plea of vested rights set up, it was shown that Parliament, in granting certain powers to the old navigators, had regard mainly to the convenience and advantage of the public; and they were not precluded from empowering a new navigation to be formed if it could be proved to present a more convenient and advantageous mode of conveyance.

On these grounds the Duke was strongly supported by the inhabitants of the localities proposed to be served by the intended canal. The "Junto of Old Navigators of the Mersey and Irwell Company" had for many years carried things with a very high hand, extorted the highest rates, and, in cases of loss by delay or damage to goods in transit, refused all redress. A feeling very hostile to them and their monopoly had accordingly grown up, which now exhibited itself in a powerful array of petitions to Parliament in favour of the Duke's bill.

On the 17th of February, 1762, the bill came before the Committee of the House of Commons, and Brindley was examined in its support. We regret that no copy of his evidence now exists * from which we might have formed an opinion of the engineer's abilities as a witness. Some curious anecdotes have, however, been

---

* Search has been made at the Bridgewater Estate Offices at Manchester, and in the archives of the Houses of Parliament, but no copy can be found. It is probable that the Parliamentary papers connected with this application to Parliament were destroyed by the fire which consumed so many similar documents about twenty-five years ago.

preserved of his demeanour and evidence on canal bills
before Parliament.   When asked, on one occasion, to
produce a drawing of an intended bridge, he replied that
he had no plan of it on paper, but he would illustrate
it by a model.   He went out and bought a *large cheese*,
which he brought into the room and cut into two equal
parts, saying, "Here is my model."   The two halves
of the cheese represented the semicircular arches of his
bridge ; and by laying over them some long rectangular
object, he could thus readily communicate to the Committee the position of the river flowing underneath and the
canal passing over it.*

On another occasion, when giving his evidence, he spoke
so frequently about "puddling," describing its uses and
advantages, that some of the members expressed a desire
to know what this extraordinary mixture was, that could
be applied to such important purposes.   Preferring a
practical illustration to a verbal description, Brindley
caused a mass of clay to be brought into the committee-
room, and, moulding it in its raw untempered state into
the form of a trough, he poured into it some water, which
speedily ran through and disappeared.   He then worked
the clay up with water to imitate the process of puddling,
and again forming it into a trough, filled it with water,
which was now held in without a particle of leakage.
"Thus it is," said Brindley, "that I form a water-tight
trunk to carry water over rivers and valleys, wherever
they cross the path of the canal." †

Again, when Brindley was giving evidence before a
Committee of the House of Peers as to the lockage of his
proposed canal, one of their Lordships asked him, "But
what is a lock?" on which the engineer took a piece of
chalk from his pocket and proceeded to explain it by

---

* Stated by Mr. Hughes, in his
'Memoir of Brindley,' as having
been communicated to him by
James Loch, Esq., M.P., formerly

agent for the Duke's Trustees.
† Memoir of Brindley,' by S.
Hughes, C.E., in 'Weale's Papers
on Civil Engineering.'

means of a diagram which he drew upon the floor, and made the matter clear at once.* He used to be so ready

---

* As the reader may possibly desire information on the same point, we may here briefly explain the nature of a Canal Lock. It is employed as a means of carrying navigations through an uneven country, and raising the boats from one water level to another, or *vice versâ*. The lock is a chamber formed of masonry, occupying the bed of the canal where the difference of level is to be overcome. It is provided with two pairs of gates, one at each end; and the chamber is so contrived that the level of the water which it contains may be made to coincide with either the higher level above, or the lower level below it. The following diagrams will explain the form and construction of the lock. A represents what is called the upper pond, B the lower, C is the left wall, and DD side culverts. When the gates at the lower end of the chamber (E) are opened, and those at the upper end (F) are closed, the water in the chamber will stand at the lower level of the canal; but when the lower gates are closed, and the upper gates are opened, the water will naturally coincide with that in the upper part of the canal. In the first case, a boat may be floated into the lock from the lower part, and then, if the lower gates be closed and water is admitted from the upper level, the canal-boat is raised, by the depth of water thus added to the lock, to the upper level, and on the complete opening of the gates it is thus floated onward. By reversing the process, it will readily be understood how the boat may, in like manner, be lowered from the higher

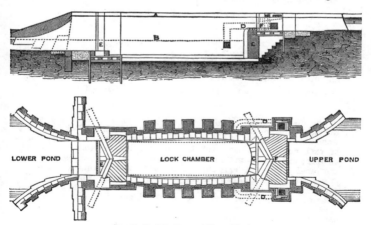

Longitudinal Section and Plan of Lock.

with his chalk for purposes of illustration, that it became a common saying in Lancashire, that "Brindley and chalk would go through the world." He was never so eloquent as when with his chalk in hand,—it stood him in lieu of tongue.

On the day following Brindley's examination before the Committee on the Duke's bill, that is, on the 18th of February, we find him entering in his note-book that the Duke sent out "200 leators" to members—possible friends of the measure; containing his statement of reasons in favour of the bill. On the 20th Mr. Tomkinson, the Duke's solicitor, was under examination for four hours and a half. Sunday intervened, on which day Brindley records that he was "at Lord Harrington's." On the following day, the 22nd, the evidence for the bill was finished, and the Duke followed this up by sending out 250 more letters to members, with an abstract of the evidence given in favour of the measure. On the 26th there was a debate of eight hours on the bill, followed by a division, in Committee of the whole House, thus recorded by Brindley :—

> "ad a grate Division of 127 fort Duk
> 98 nos
> —
> for t⁰ Duk 29 Me Jorete "

But the bill had still other discussions and divisions to encounter before it was safe. The Duke and his agents worked with great assiduity. On the 3rd of March he

---

to the lower level. The greater the lift or the lowering, the more water is consumed in the process of exchange from one level to another; and where the traffic of the canal is great, a large supply of water is required to carry it on, which is usually provided by capacious reservoirs situated above the summit level. Various expedients are adopted for economising water: thus, when the width of the canal will admit of it, the lock is made in two compartments, communicating with each other by a valve, which can be opened and shut at pleasure; and by this means one-half of the water which it would otherwise be necessary to discharge to the lower level may be transferred to the other compartment.

caused 250 more letters to be distributed amongst the members ; and on the day after we find the House wholly occupied with the bill. We quote again from Brindley's record : " 4 [March] ade bate at the Hous with grate vigor 3 divisons the Duke carred by Numbers evory time a 4 division moved but Noos yelded." On the next day we read " wont thro the closos ; " from which we learn that the clauses were settled and passed. Mr. Gilbert and Mr. Tomkinson then set out for Lancashire : the bill was safe. It passed the third reading, Brindley making mention that " Lord Strange " was " sick with geef [grief] on that affair Mr. Wellbron want Rong god,"—which latter expression we do not clearly understand, unless it was that Mr. Wilbraham wanted to wrong God. The bill was carried to the Lords, Brindley on the 10th March making the entry, " Touk the Lords oath." But the bill passed the Upper House " without opposishin," and received the Royal Assent on the 24th of the same month.

On the day following the passage of the bill through the House of Lords (of which Brindley makes the triumphant entry, " Lord Strange defetted "), he set out for Lancashire, after nine weary weeks' stay in London. To hang about the lobbies of the House and haunt the office of the Parliamentary agent, must have been excessively irksome to a man like Brindley, accustomed to incessant occupation and to see works growing under his hands. During this time we find him frequently at the office of the Duke's solicitor in " Mary Axs ; " sometimes with Mr. Tomkinson, who paid him his guinea a-week during the latter part of his stay ; and on several occasions he is engaged with gentlemen from the country, advising them about " saltworks at Droitwitch " and mill-arrangements in Cheshire.

Many things had fallen behind during his absence and required his attention, so he at once set out home ; but the first day, on reaching Dunstable, he was alarmed to find that his mare, so long unaccustomed to the road, had "allmost lost ye use of her Limes" [limbs]. He therefore went on slowly, as the mare was a great favourite with him—his affection for the animal having on one occasion given rise to a serious quarrel between him and Mr. Gilbert —and he did not reach Congleton until the sixth day after his setting out from London. He rested at Congleton for two days, during which he "settled the geering of the silk-mill," and then proceeded straight on to Worsley to set about the working survey of the new canal.

The course of this important canal, which unites the mills of Manchester with the shipping of Liverpool, is about twenty-four miles in length.* From Longford Bridge, near Manchester, its course lies in a south-westerly direction for some distance, crossing the river Mersey at a point about five miles above its junction with the Irwell. At Altrincham it proceeds in a westerly direction, crossing the river Bollin about three miles further on, near Dunham. After crossing the Bollin, it describes a small semi-circle, proceeding onward in the valley of the Mersey, and nearly in the direction of the river as far as the crossing of the high road from Chester to Warrington. It then bends to the south to preserve the high level, passing in a southerly direction as far as Preston, in Cheshire, from whence it again turns round to the north to join the river Mersey. [For Map of the Canal, see pp. 168–9.]

The canal lies entirely in the lower part of the new red sandstone, the principal earthworks consisting of the clays, marls, bog-earths, and occasionally the sandstones of this formation. The heaviest bog crossed in the line of the works was Sale Moor, west of the Mersey, where the bottom was of quicksand; and the construction of the

---

* The following statement of the lengths of the different portions of the Duke's canal, including those originally executed, is from the map published by Brindley in 1769 :—

	Miles.	furl.	chains.	
From Worsley to Longford Bridge	6	0	0	Level.
„ Longford Bridge to Manchester	4	2	0	„
„ Longford Bridge to Preston Brook	19	0	0	.,
„ Preston Brook to upper part of Runcorn ..	4	4	0	„
„ Upper part of Runcorn to the Mersey	0	5	7	79 feet fall.

canal at that part was probably an undertaking of as
formidable a character as the laying of the railroad over
Chat Moss proved some sixty years later. But Brindley,
like Stephenson, looked upon a difficulty as a thing to be
overcome; and when an obstruction presented itself, he at
once set his wits to work and studied how it was best to
be grappled with and surmounted. A large number of
brooks had to be crossed, and also two important rivers,
which involved the construction of numerous aqueducts,
bridges, and culverts, to provide for the surface water
supply of the district. It will, therefore, be obvious that
the undertaking was of a much more important nature
—more difficult for the engineer to execute, and more
costly to the noble proprietor who found the means for
carrying it to a completion — than the comparatively
limited and inexpensive work between Worsley and Manchester, which we have above described.

The capital idea which Brindley early formed and determined to carry out, was to construct a level of dead water
all the way from Manchester to a point as near to the
junction of the canal with the Mersey as might be found
practicable. Such a canal, he clearly saw, would not be
so expensive to work as one furnished with locks at intermediate points. Brindley's practice of securing long levels
of water in canals was in many respects similar to that of
George Stephenson with reference to flat gradients upon
railways; and in all the canals that he constructed, he
planned and carried them out as far as possible after this
leading principle. Hence the whole of the locks on the
Duke's canal were concentrated at its lower end near
Runcorn, where the navigation descended, as it were by a
flight of water steps, into the river Mersey. Lord Ellesmere has observed that this uninterrupted level of the
Bridgewater Canal from Leigh and Manchester to Runcorn,
and the concentration of its descent to the Mersey at the
latter place, have always been considered as among the
most striking evidences of the genius and skill of Brindley.

There was, as usual, considerable delay in obtaining possession of the land on which to commence the works.   The
tenants required a certain notice, which must necessarily
expire before the Duke's engineer could take possession;
and numerous obstacles were thrown in his way, both by
tenants and landlords hostile to the undertaking.   In many
cases the Duke had to pay dearly for the land purchased
under the compulsory powers of his Act.   Near Lymm, the
canal passed through a little bit of garden belonging to a
poor man's cottage, the only produce growing upon the
ground being a pear-tree.   For this the Duke had to pay
thirty guineas, and it was thought a very extravagant price
at that time.   Since the introduction of railways, the price
would probably be considered ridiculously low.   For the land
on which the warehouses and docks were built at Manchester, the Duke had to pay in all the much more formidable
sum of about forty thousand pounds.

The Old Quay Navigation, even at the last moment,
thought to delay if not to defeat the Duke's operations, by
lowering their rates nearly one-half.   Only a few days after
the Royal Assent had been given to the bill, they published
an announcement, appropriately dated the 1st of April,
setting forth the large sacrifices they were about to make,
and intimating that "from their Reductions in Carriage a
real and permanent Advantage will arise to the Public,
and they will experience that Utility so cried up of late,
but which has hitherto only existed in promises."   The
Duke heeded not the ineffective blow thus aimed at him:
he was only more than ever resolved to go forward with
his canal.   He was even offered the Mersey Navigation
itself at the price of thirteen thousand pounds; but he
would not have it now at any price.

The public spirit and enterprise displayed by many of
the young noblemen of those days was truly admirable.
Brindley had for several years been in close personal communication with Earl Gower as to the construction of the
canal intended to unite the Mersey with the Trent and the

Severn, and thus connect the ports of Liverpool, Hull, and
Bristol, by a system of inland water-communication.   With
this object, as we have seen, he had often visited the Earl
at his seat at Trentham, and discussed with him the plans
by which this truly magnificent enterprise was to be carried
out; and he had frequently visited the Earl of Stamford
at his seat at Enville for the same purpose.   But those
schemes were too extensive and costly to be carried out
by the private means of either of those noblemen, or even
by both combined.   They were, therefore, under the
necessity of stirring up the latent enterprise of the landed
proprietors in their respective districts, and waiting until
they had received a sufficient amount of local support to
enable them to act with vigour in carrying their great
design into effect.

The Duke of Bridgewater's scheme of uniting Manchester
and Liverpool by an entirely new line of water-communication, cut across bogs and out of the solid earth in some
places, and carried over rivers and valleys at others by
bridges and embankments, was scarcely less bold or costly.
Though it was spoken of as another of the Duke's "castles
in the air," and his resources were by no means overflowing
at the time he projected it, he nevertheless determined to
go on alone with it, should no one be willing to join him.
The Duke thus proved himself a real Dux or leader of
industrial enterprise in his district; and by cutting his
canal, and providing a new, short, and cheap water-way
between Liverpool and Manchester, which was afterwards
extended through the counties of Chester, Stafford, and
Warwick, he unquestionably paved the way for the creation
and development of the modern manufacturing system existing in the north-western counties of England.

We need scarcely say how admirably he was supported
throughout by the skill and indefatigable energy of his
engineer.   Brindley's fertility in resources was the theme
of general admiration.   Arthur Young, who visited the
works during their progress, speaks with enthusiastic

admiration of his "bold and decisive strokes of genius," his "penetration which sees into futurity, and prevents obstructions unthought of by the vulgar mind, merely by foreseeing them : a man," says he, "with such ideas, moves in a sphere that is to the rest of the world imaginary, or at best a *terra incognita.*"

It would be uninteresting to describe the works of the Bridgewater Canal in detail; for one part of a canal is usually so like another, that to do so were merely to involve a needless amount of repetition of a necessarily dry description. We shall accordingly content ourselves with referring to the original methods by which Brindley contrived to overcome the more important difficulties of the undertaking.

From Longford Bridge, where the new works commenced, the canal, which was originally about eight yards wide and four feet deep, was carried upon an embankment of about a mile in extent across the valley of the Mersey. One might naturally suppose that the conveyance of such a mass of earth must have exclusively employed all the horses and carts in the neighbourhood for years. But Brindley, with his usual fertility in expedients, contrived to make the construction of one part of the canal subservient to the completion of the remainder. He had the stuff required to make up the embankment brought in boats partly from Worsley and partly from other parts of the canal where the cutting was in excess; and the boats, filled with this stuff, were conducted from the canal along which they had come into watertight caissons or cisterns placed at the point over which the earth and clay had to be deposited.

The boats, being double, fixed within two feet of each other, had a triangular trough supported between them of sufficient capacity to contain about seventeen tons of earth. The bottom of this trough consisted of a line of trap-doors, which flew open at once on a pin being drawn, and discharged their whole burthen into the bed of the canal in an instant. Thus the level of the embankment was raised

to the point necessary to enable the canal to be carried
forward to the next length. Arthur Young was of opinion
that the saving effected by constructing the Stretford
embankment in this way, instead of by carting the stuff,
was equivalent to not less than five thousand per cent. !
The materials of the caissons employed in executing this
part of the work were afterwards used in forming temporary locks across the valley of the Bollin, whilst the
embankment was being constructed at that point by a
process almost the very reverse, but of equal ingenuity.

Brindley's Ballast Boats.

In the same valley of the Mersey the canal had to be
carried over a large brook subject to heavy floods, by means
of a strong bridge of two arches, adjoining which was a
third, affording provision for a road. Further on, the canal
was carried over the Mersey itself upon a bridge with one
arch of seventy feet span. Westward of this river lay a
very difficult part of the work, occasioned by the carrying
of the navigation over the Sale Moor Moss. Many thought
this an altogether impracticable thing ; as not only had the
hollow trunk of earth in which the canal lay to be made
water-tight, but to preserve the level of the water-way it
must necessarily be raised considerably above the level of
the Moor across which it was to be laid. Brindley overcame
the difficulty in the following manner. He made a strong
casing of timber-work outside the intended line of embankment on either side of the canal, by placing deal balks in
an erect position, backing and supporting them on the outside with other balks laid in rows, and fast screwed together ; and on the front side of this woodwork he had his

earth-work brought forward, hard rammed, and puddled, to form the navigable canal; after which the casing was moved onward to the part of the work further in advance. and the bottom having previously been set with rubble and gravel, the embankment was thus carried forward by degrees, the canal was raised to the proper level, and the whole was substantially and satisfactorily finished.

A steam-engine of Brindley's contrivance was erected at Dunham Town Bridge to pump the water from the foundations there. The engine was called a Sawney, for what reason is not stated, and, for long after, the bridge was called Sawney's Bridge. The foundations of the under-bridge, near the same place, were popularly supposed to be set on quicksand; and old Lord Warrington, when he had occasion to pass under it, would pretend cautiously to look about him, as if to examine whether the piers were all right, and then run through as fast as he could. A tall poplar-tree stood at Dunham Banks, on which a board was nailed showing the height of the canal level; the people long after called the place "The Duke's Folly," the name given to it while his scheme was still believed to be impracticable. But the skill of the engineer baffled these and other prophets of evil; and the success of his expedients, in nearly every case of difficulty that occurred, must certainly be regarded as remarkable, considering the novel and unprecedented character of the undertaking.

Brindley invariably contrived to economise labour as much as possible, and many of his expedients with this object were very ingenious. So far as he could, he endeavoured to make use of the canal itself for the purpose of forwarding the work. He had a floating blacksmith's forge and shop, provided with all requisite appliances, fitted up in one barge; a complete carpenter's shop in another; and a mason's shop in a third; all of which were floated on as the canal advanced, and were thus always at hand to supply the requisite facilities for prosecuting the operations with economy and despatch. Where there was a

break in the line of work, occasioned, for instance, by the
erection of some bridge not yet finished, the engineer had
similar barges constructed and carried by land to other
lengths of the canal which were in progress, where they
were floated and advanced in like manner for the use of the
workmen.   When the bridge across the Mersey, which was
pushed on as rapidly as possible with the object of economising labour and cost of materials, was completed, the
stone, lime, and timber were brought along the canal from
the Duke's property at Worsley, as well as supplies of clay
for the purpose of puddling the bottom of the waterway;
and thus the work rapidly advanced at all points.

As one of the great objections made to the construction
of the canal had been the danger threatened to the surrounding districts by the bursting of the embankments,
Brindley made it his object to provide against the occurrence of such an accident by an ingenious expedient.   He
had stops or flood-gates contrived and laid in various parts
of the bed of the canal, across its bottom, so that, in the
event of a breach occurring in the bank and a rush of
water taking place, the current which must necessarily
set in to that point should have the effect of immediately
raising the valvular floodgates, and so shutting off the
stream and preventing the escape of more water than was
contained in the division between the two nearest gates
on either side of the breach.   At the same time, these
floodgates might be used for cutting off the waters of the
canal at different points, for the purpose of making any
necessary repairs in particular lengths; the contrivance of
waste tubes and plugs being so arranged that the bed of
any part of the canal, more especially where it passed over
the bridges, might be laid bare in a few hours, and the
repairs executed at once.

In devising these ingenious expedients, it ought to be
remembered that Brindley had no previous experience to
fall back upon, and possessed no knowledge of the means
which foreign engineers might have adopted to meet

I.   P similar emergencies. All had been the result of his own original thinking and contrivance; and, indeed, many of these devices were altogether new and original, and had never before been tried by any engineer.

It is curious to trace the progress of the works by Brindley's own memoranda, which, though brief, clearly exhibit his marvellous industry and close application to every detail of the business. He settled with the farmers for their tenant-right, sold and accounted for the wood cut down and the gravel dug out along the line of the canal, paid the workmen employed,* laid out the work, measured off the quantities done from time to time, planned and erected the bridges, designed the canal boats required for conveying the earth to form the embankments, and united in himself the varied functions of land-surveyor, carpenter, mason, brickmaker, boat-builder, paymaster, and engineer. We even find him condescending to count bricks and sell grass. Nothing was too small for him to attend to, nor too bold for him to undertake, when necessity required. At the same time we find him contriving a water-plane for the Duke's collieries at Worsley, and occasionally visiting Newchapel, Leek, and Congleton, in Cheshire, for the purpose of attending to the business on which he still continued to be employed at those places.

The heavy works at the crossing of the Mersey occupied

---

* The following bill is preserved amongst the Bridgewater Canal papers. Simcox was a skilled mechanic, and acted as foreman of the carpenters :—

"His Grace the Duke of Bridgewater to Sam¹ Simcox.　Dʳ

		£.	s.	d.
23 Marʰ 1760	To 12 days work at 21ᵈ per　..　..　..　..	1	1	0
23 Augᵗ	To 6 days more dᵒ at dᵒ ..　..　..　..　..	0	10	6
6 Sepʳ	To 8 days more dᵒ at dᵒ ..　..　..　..　..	0	14	0
		2	5	6

1 Novʳ 1760. Recᵈ the Contents above by the Hands of John Gilbert for the Use of Sam¹ Simcox.　　　　　　　　Pᵈ
　　　　　　　　　　　　　　　　　　　　　JAMES BRINDLEY."

The wages of what was called a "right-hand man" at that time were from 14d. to 16d. a day, and of a "left-hand man" from 1s. to 14d.

him almost exclusively towards the end of the year 1763.
He was there making dams and pushing on the building of
the bridge. Occasionally he enters the words, " short of
men at Cornbrook." Indeed, he seems at that time to have
lived upon the works, for we find the almost daily entry of
" dined at the Bull, 8d." On the 10th of November he
makes this entry : " Aftor noon sattled about the size of the
arch over the river Marsee [Mersey] to be 66 foot span and
rise 16·4 feet." Next day he is " landing balk out of the
ould river in to the canal." Then he goes on, " I proseeded
to Worsley Mug was corking ye boats the mascns woss
making the senter of the waire [weir]. Whith⁵ was osing
to put the lator side of the water-wheel srouds on I orderd
the pit for ye spindle of ye morter-mill to be sunk level
with ye canal Mr. Gilbert sade ye 20 Tun Boat should be
at ye water mitang [meeting] by 7 o'clock the next morn."
Next morning he is on the works at Cornhill, setting " a
carpentar to make scrwos " [screws], superintending the
gravelling of the towing-path, and arranging with a farmer
as to Mr. Gilbert's slack. And so he goes on from day to
day with the minutest details of the undertaking.

He was not without his petty " werrets " and troubles
either. Brindley and Gilbert do not seem to have got on
very well together. They were both men of strong tem-
pers, and neither would tolerate the other's interference.
Gilbert, being the Duke's factotum, was accustomed to call
Brindley's men from their work, which the other would not
brook. Hence we have this entry on one occasion,—" A
meshender [messenger] from Mr G I retorned the anser
No more sosiety." In fact, they seem to have quarrelled.*

---

* The Earl of Bridgewater, in
his rambling ' Letter to the Parisians,' above referred to, alleges
that the quarrel originated in Gilbert's horse breaking into the field
where Brindley's mare was grazing,
and doing her such injury that the
engineer was for a time prevented
using the animal in the pursuit of
his business. The mare was a
great favourite of Brindley's, and
he is said to have taken the thing
very much to heart. The Earl alleges that Brindley was under the
impression Gilbert had contrived
the trick out of spite.

We find the following further entries on the subject in Brindley's note-book: "Thursday 17 Novr past 7 o'clock at night M Gilbert and sun Tom caled on mee at Gorshill and I went with them to ye Coik [sign of the Cock] tha stade all night and the had balk [blank?] bill of parsill 18 Fryday November 7 morn I went to the Cock and Bruck-fast with Gilberts he in davred to imploye ye carpinters at Cornhill in making door and window frames for a Building in Castle field and shades for the mynors in Dito and other things I want them to Saill Moor Hee took upon him diriction of ye back drains and likwaise such Lands as be twixt the 2 hous and ceep uper side the large farme and was displesed with such raing as I had pointed out."

Those differences between Brindley and Gilbert were eventually reconciled, most probably by the mediation of the Duke, for the services of both were alike essential to him; and we afterwards find them working cordially together and consulting each other as before on any important part of the undertaking.

During the construction of Longford Bridge, Brindley seems, from his note-book, to have entertained considerable apprehensions as to its ability to resist the heavy floods with which it was threatened. Thus, on the 26th of November, 1763, he enters:—"Grate Rains the canal rose 2 inches extra I dreed fr [4?] clock at Longfoard;" and on the following day, which was a Sunday, he writes:—"Lay in Bad till noon floode and Raine." Then in the afternoon he adds, "The water in Longfoord Brook was withe in six inches of the high of the santer [centre] of ye waire [weir?]." The bridge, however, stood firm; and when the flood subsided, the building was again proceeded with; and by the end of the year it was finished and gravelled over, while the embankment was steadily proceeding beyond the Mersey in the manner above described.

Brindley did not want for good workmen to carry out his plans. He found plenty of labourers in the neighbourhood accustomed to hard work, who speedily became

expert excavators; and though there was at first a lack
of skilled carpenters, blacksmiths, and bricklayers, they
soon became trained into such under the vigilant eye of

Longford Bridge.

so able a master as Brindley was. We find him, in his
note-book, often referring to the men by their names,
or rather byenames; for in Lancashire proper names
seem to have been little used at that time. "Black
David" was one of the foremen most employed on
difficult matters, and "Bill o Toms" and "Busick Jack"
seem also to have been confidential workmen in their
respective departments. We are informed by a gentleman of the neighbourhood * that most of the labourers
employed were of a superior class, and some of them
were "wise" or "cunning men," blood-stoppers, herb-
doctors, and planet-rulers, such as are still to be found
in the neighbourhood of Manchester. Their very super-
stitions, says our informant, made them thinkers and
calculators. The foreman bricklayer, for instance, as
his son used afterwards to relate, always "ruled the planets

---

* R. Rawlinson. Esq., C.E., Engineer to the Bridgewater Canal.

to find out the lucky days on which to commence any important work," and he added, "none of our work ever gave way." The skilled men had their trade-secrets, in which the unskilled were duly initiated,—simple matters in themselves, but not without their uses. The following may be taken as specimens of the secrets of embanking in those days:—

A wet embankment can be prevented from slipping by dredging or dusting powdered lime in layers over the wet clay or earth.

Sand or gravel can be made water-tight by shaking it together with flat bars of iron run in some depth, say two feet, and washing down loam or soil as the bars are moved about, thus obviating the necessity for clay puddle.

Dry-rot can be prevented in warehouses by setting the bricks opposite the ends of the main beams of the warehouse in dry sand.

Whilst constructing the canal, Brindley was very intimate with one Lawrence Earnshaw, of Mottram-in-Longdendale, a kindred mechanical genius, though in a smaller way. Lawrence was a very poor man's son, and had served a seven years' apprenticeship to the trade of a tailor, after which he bound himself apprentice to a clothier for seven years ; but these trades not suiting his tastes, and being of a decidedly mechanical turn, he finally bound himself apprentice to a clockmaker, whom he also served for seven years. This eccentric person invented many curious and ingenious machines, which were regarded as of great merit in his time. One of these was an astronomical and geographical machine, beautifully executed, showing the earth's diurnal and annual motion, after the manner of an orrery. The whole of the calculations were made by himself, and the machine is said to have been so exactly contrived and executed that, provided the vibration of the pendulum did not vary, the machine would not alter a minute in a hundred years; but this might probably be an extravagant estimate on the part of Earnshaw's friends. He was also a musical instrument maker and music teacher, a worker in metals and in wood, a painter and glazier, an optician, a bellfounder, a chemist and metallurgist, an engraver—in short, an almost universal mechanical genius. But though he could make all these things, it is mentioned as a remarkable fact, that with all his ingenuity, and after many efforts (for he made many), he never could make a wicker-basket! Indeed, trying to be a universal genius was his ruin. He did, or attempted to do, so much, that he never stood still and established himself in any one thing; and, notwithstanding his great ability, he died "not worth a groat." Amongst Earnshaw's various contrivances was a piece of machinery to raise water from a coal-mine at Hague, near Mottram, and (about 1753) a machine to spin and reel cotton at one operation—in fact, a spinning-jenny—which he showed to some of his neighbours as a curiosity, but, after having convinced them of what might be done by its means, he immediately destroyed it, saying that "he would not be the means of taking bread out of the mouths of the poor." * He was a total abstainer from strong drink, long before the days of Teetotal Societies. Towards the end of his life he continued on intimate terms with Brindley; and when they met they did not readily separate.

While the undertaking was in full progress, from four to six hundred men were employed; they were divided into gangs of about fifty, each of which was directed by a captain and setter-out of the works. One who visited the canal during its construction in 1765, wrote thus of the busy scene which the works presented: "I surveyed

---

* Mr. Bennet Woodcroft, of the Patent Office, writes us as follows with reference to Earnshaw's alleged invention of a spinning machine :—"The fact really is, that the machine in question was invented by John Kay of Bury; and when, in 1753, a mob broke into Kay's house, and completely gutted it, the model of the spinning machine was saved by Earnshaw who subsequently destroyed it."

the Duke's men for two hours, and think the industry
of bees or labour of ants is not to be compared to them.
Each man's work seems to depend on and be connected
with his neighbour's, and the whole posse appeared as I
conceive did that of the Tyrians when they wanted houses
to put their heads in at Carthage." * At Stretford the
visitor found "four hundred men at work, putting the
finishing stroke to about two hundred yards of the canal,
which reached nearly to the Mersey, and which, on drawing up the floodgates, was to receive a proper quantity
of water and a number of loaded barges. One of these
appeared like the hull of a collier, with its deck all covered,
after the manner of a cabin, and having an iron chimney
in the centre; this, on inquiry, proved to be the carpentry,
but was shut up, being Sabbath-day, as was another barge,
which contained the smith's forge. Some vessels were
loaded with soil, which was put into troughs (see cut, p.
205), fastened together, and rested on boards that lay
across two barges; between each of these there was room
enough to discharge the loading by loosening some iron
pins at the bottom of the troughs. Other barges lay
loaded with the foundation-stones of the canal bridge,
which is to carry the navigation across the Mersey. Near
two thousand oak piles are already driven to strengthen
the foundations of this bridge. The carpenters on the
Lancashire side were preparing the centre frame, and on
the Cheshire side all hands were at work in bringing down
the soil and beating the ground adjacent to the foundations of the bridge, which is designed to be covered
with stone in a month, and finished in about ten days
more." †

By these vigorous measures the works proceeded rapidly
towards completion. Before, however, they had made any

---

* St. James's Chronicle, July
1st, 1765.
† 'A History of Inland Navigations. Particularly those of the
Duke of Bridgewater in Lancashire and Cheshire.' 2nd Ed., p. 39.

progress at the Liverpool end, Earl Gower, encouraged and assisted by the Duke, had applied for and obtained an Act to enable a line of navigation to be formed between the Mersey and the Trent; the Duke agreeing with the promoters of the undertaking to vary the course of his canal and meet theirs about midway between Preston-brook and Runcorn, from which point it was to be carried northward towards the Mersey, descending into that river by a flight of ten locks, the total fall being not less than 79 feet from the level of the canal to low-water of spring-tides.

When this deviation was proposed, the bold imagination of Brindley projected a bridge across the tideway of the Mersey itself, which was there some four hundred and sixty yards wide, with the object of carrying the Duke's navigation directly onward to the port of Liverpool on the Lancashire side of the river.*   This was an admirable idea, which, if carried out, would probably have redounded more to the fame of Brindley than any other of his works. But the cost of that portion of the canal which had already been executed, had reached so excessive an amount, that the Duke was compelled to stop short at Runcorn, at which place a dock was constructed for the accommodation of the shipping employed in the trade connected with the undertaking.

From Runcorn, it was arranged that the boats should navigate by the open tideway of the Mersey to the harbour of Liverpool, at which place the Duke made arrangements to provide another dock for their accommodation. Brindley

---

* This bold scheme, so earnestly advocated by Brindley, was thus noticed by a Liverpool paper of the time :—" On Monday last, Mr. Brindley waited upon several of the principal gentlemen of this town and otLers at Runcorn, in order to ascertain the expense that may attend the building of a bridge over the river Mersey at the latter place, which is estimated at a sum inferior to the advantages that must arise both to the counties of Lancaster and Chester from a communication of this sort."—Williamson's ' Liverpool Advertiser,' July 19th, 1768.

made frequent visits to Liverpool for the purpose of directing its excavation, and he superintended it until its completion. The Duke's Dock lies between the Salthouse and Albert Docks on the north, and the Wapping and King's Docks on the south. The Salthouse was the only public dock near it at the time that Brindley excavated this basin. There were only three others in Liverpool to the north, and not one to the south; but the Duke's Dock is now the centre of about five miles of docks, extending from it on either side along the Lancashire shore of the Mersey; and it continues to this day to be devoted to the purposes of the navigation.

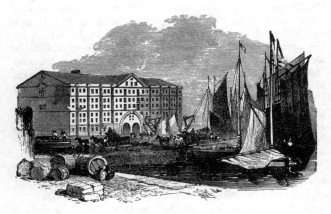

The Duke's Dock, Liverpool.

CHAPTER V.

THE DUKE'S DIFFICULTIES — COMPLETION OF THE CANAL —
GROWTH OF MANCHESTER.

LONG before the Runcorn locks were constructed, and the
canal from Longford Bridge to the Mersey was available
for purposes of traffic, the Duke found himself reduced to
the greatest straits for want of money. Numerous unexpected difficulties had occurred, so that the cost of the
works considerably exceeded his calculations; and though
the engineer carried on the whole operations with the
strictest regard to economy, the expense was nevertheless
almost more than any single purse could bear. The execution of the original canal from Worsley to Manchester
cost about a thousand guineas a mile, besides the outlay
upon the terminus at Manchester. There was also the
expenditure incurred in building the requisite boats for the
canal, in opening out the underground workings of the
collieries at Worsley, and in erecting various mills, workshops, and warehouses for carrying on the new business.

The Duke was enabled to do all this without severely
taxing his resources, and he even entertained the hope
of being able to grapple with the still greater undertaking
of cutting the twenty-four miles of new canal from Longford Bridge to the Mersey. But before these works were
half finished, and whilst the large amount of capital
invested in them was lying entirely unproductive, he
found that the difficulties of the undertaking were likely
to prove too much for him. Indeed, it seemed an enterprise beyond the means of any private person, and more like
that of a monarch with State revenues at his command,
than of a young English nobleman with only his private
resources.

But the Duke was possessed by a brave spirit. He had put his hand to the work, and he would not look back. He had become thoroughly inspired by his great idea, and determined to bend his whole energies to the task of carrying it out. He was only thirty years of age—the owner of several fine mansions in different parts of the country, surrounded by noble domains—he had a fortune sufficiently ample to enable him to command the pleasures and luxuries of life, so far as money can secure them; yet he voluntarily denied himself their enjoyment, and chose to devote his time to consultations with an unlettered engineer, and his whole resources to the cutting of a canal to unite Liverpool and Manchester.

Taking up his residence at the Old Hall at Worsley— a fine specimen of the old timbered houses so common

Worsley Old Hall.

in South Lancashire and the neighbouring counties,—he cut down every unnecessary personal expense; denied himself every superfluity, except perhaps that of a pipe of tobacco; paid off his retinue of servants; put down his carriages and town house; and confined himself and his Ducal establishment to a total expenditure of 400*l.* a-year. A horse was, however, a necessity, for the

purpose of enabling him to visit the canal works during
their progress at distant points; and he accordingly continued to maintain one horse for himself and another for
his groom.

Notwithstanding this rigid economy, the Duke still found
his resources inadequate to meet the heavy cost of vigorously carrying on the undertaking, and on Saturday nights
he was often put to the greatest shifts to raise the requisite money to pay his large staff of craftsmen and labourers.
Sometimes their payment had to be postponed for a week
or more, until the cash could be raised by sending round
for contributions among the Duke's tenantry. Indeed,
his credit fell to the lowest ebb, and at one time he could
not get a bill for 500*l.* cashed in either Liverpool or Manchester.*

He was under the necessity of postponing all payments
that could be avoided, and it went abroad that the Duke
was "drowned in debt." He tried to shirk even the payment of his tithes, and turned a deaf ear to all the applications of the collector. At length the rector himself
determined to waylay him. But the Duke no sooner
caught sight of him coming across his path than he bolted!
The rector was not thus to be baulked. He followed—pursued—and fairly ran his debtor to earth in a saw-pit! The
Duke was not a little amused at being hunted in such a
style by his parson, and so soon as he found his breath,
he promised payment, which shortly followed.

When Mr. George Rennie, the engineer, was engaged,
in 1825, in making the revised survey of the Liverpool
and Manchester Railway, he lunched one day at Worsley
Hall with Mr. Bradshaw, manager of the Duke's property,

---

* There is now to be seen at
Worsley, in the hands of a private
person, a promissory note given by
the Duke, bearing interest, for as
low a sum as five pounds. Amongst
the persons known to be lenders of
money, to whom the Duke applied
at the time, was Mr. C. Smith, a
merchant at Rochdale; but he
would not lend a farthing, believing the Duke to be engaged
in a perfectly ruinous undertaking.

then a very old man. He had been a contemporary of the Duke, and knew of the monetary straits to which his Grace had been reduced during the construction of the works. Whilst at table, Mr. Bradshaw pointed to a small whitewashed cottage on the Moss, about a mile and a half distant, and said that in that cottage, formerly a public-house, the Duke, Brindley, and Gilbert had spent many an evening discussing the prospects of the canal while in progress. One of the principal topics of conversation on those occasions was the means of raising funds against the next pay night. "One evening in particular," said Mr. Bradshaw, " the party was unusually dull and silent. The Duke's ready-money was exhausted; the canal was not nearly finished; his Grace's credit was at the lowest ebb; and he was at a loss what step to take next. There they sat, in the small parlour of the little public-house, smoking their pipes, with a pitcher of ale before them, melancholy and silent. At last the Duke broke the silence by asking in a querulous tone, 'Well, Brindley, what's to be done now? How are we to get at the money for finishing this canal?' Brindley, after a few long puffs, answered through the smoke, 'Well, Duke, I can't tell; I only know that if the money can be got, I can finish the canal, and that it will pay well.' 'Ay,' rejoined the Duke, 'but *where* are we to get the money?' Brindley could only repeat what he had already said; and thus the little party remained in moody silence for some time longer, when Brindley suddenly started up and said, 'Don't mind, Duke; don't be cast down; we are sure to succeed after all!' The party shortly after separated, the Duke going over to Worsley to bed, to revolve in his mind the best mode of raising money to complete his all-absorbing project."

One of the expedients adopted was to send Gilbert, the agent, upon a round of visits among the Duke's tenants, raising five pounds here and ten pounds there, until he had gathered together enough to pay the week's wages.

Whilst travelling about among the farmers on one of such occasions, Gilbert was joined by a stranger horseman, who entered into conversation with him; and it very shortly turned upon the merits of their respective horses. The stranger offered to swap with Gilbert, who, thinking the other's horse better than his own, agreed to the exchange. On afterwards alighting at a lonely village inn, which he had not before frequented, Gilbert was surprised to be greeted by the landlord with mysterious marks of recognition, and still more so when he was asked if he had got a good booty. It turned out that he had exchanged horses with a highwayman, who had adopted this expedient for securing a nag less notorious than the one which he had exchanged with the Duke's agent.*

At length, when the tenantry could furnish no further advances, and loans were not to be had on any terms in Manchester or Liverpool, and the works must needs come to a complete stand unless money could be raised to pay the workmen, the Duke took the road to London on horseback, attended only by his groom, to try what could be done with his London bankers. The house of Messrs. Child and Co., Temple Bar, was then the principal banking-house in the metropolis, as it is the oldest; and most of the aristocratic families kept their accounts there. The Duke had determined at the outset of his undertaking not to mortgage his landed property, and he had held to this resolution. But the time arrived when he could not avoid borrowing money of his bankers on such other security as he could offer them. He had already created a valuable and lucrative property, which was happily available for the purpose. The canal from Worsley to Manchester had proved remunerative in an extraordinary degree, and was already producing a large income. He had not the same scruples as to the pledging of the revenues of his canal that he had to the mortgaging

* The Earl of Ellesmere's 'Essays on History, Biography,' &c., p. 236.

of his lands; and an arrangement was concluded with the Messrs. Child under which they agreed to advance the Duke sums of money from time to time, by means of which he was eventually enabled to finish the entire canal.

The Messrs. Child and Co. have kindly permitted an examination of their books to be made for the purposes of this memoir; and we are accordingly enabled to state that from them it appears that the Duke obtained his first advance of 3,800*l.* from the firm about the middle of the year 1765, at which time he was in the greatest difficulty; shortly after a further sum of 15,000*l.*; then 2,000*l.*, and various other sums, making a total of 25,000*l.*; which remained owing until the year 1769, when the whole was paid off — doubtless from the profits of the canal traffic as well as the economised rental of the Duke's unburthened estates.

The entire level length of the new canal from Longford Bridge to the upper part of Runcorn, nearly twenty-eight miles in extent, was finished and opened for traffic in the year 1767, after the lapse of about five years from the passing of the Act. The formidable flight of locks, from the level part of the canal down to the waters of the Mersey at Runcorn, were not finished for several years later, by which time the receipts derived by the Duke from the sale of his coals and the local traffic of the undertaking enabled him to complete them with comparatively little difficulty. Considerable delay was occasioned by the resistance of an obstinate landowner near Runcorn, Sir Richard Brooke, who interposed every obstacle which it was in his power to offer; but his opposition too was at length overcome, and the new and complete line of water-communication between Manchester and Liverpool was finally opened throughout.

In a letter written from Runcorn, dated the 1st January, 1773, we find it stated that "yesterday the locks were opened, and the *Heart of Oak*, a vessel of 50 tons burden, for Liverpool, passed through them. This day, upwards of six hundred of his Grace's workmen were entertained upon the lock banks with an ox roasted whole and plenty of good liquor. The Duke's health and many other toasts were drunk with the loudest acclamations by the multitude, who crowded from all parts of the country to be spectators of these astonishing works. The gentlemen of the country for a long time en-

The Locks at Runcorn.

tertained a very unfavourable opinion of this undertaking, esteeming it too difficult to be accomplished, and fearing their lands would be cut and defaced without producing any real benefit to themselves or the public; but they now see with pleasure that their fears and apprehensions were ill-grounded, and they join with one voice in applauding the work, which cannot fail to produce the most beneficial consequences to the landed property, as well as to the trade and commerce of this part of the kingdom."

Whilst the canal works had been in progress, great changes had taken place at Worsley. The Duke had year by year been extending the workings of the coal; and when the King of Denmark, travelling under the title of Prince Travindahl, visited the Duke in 1768, the tunnels had already been extended for nearly two miles under the hill. When the Duke began the works, he possessed only such of the coal-mines as belonged to the Worsley estate; but he purchased by degrees the adjoining lands containing seams of coal which run under the high ground between Worsley, Bolton, and Bury; and in course of time the underground canals connecting the different workings extended for a distance of nearly forty miles. Both the hereditary and the purchased mines are worked upon two main levels, though in all there are four different levels, the highest being a hundred and twenty yards above the lowest. In opening up the underground workings the Duke is said to have expended about 168,000*l.*; but the immense revenue derived from the sale of the coals by canal rendered this an exceedingly productive outlay. Besides the extension of the canal along these tunnels, the Duke subsequently carried a branch by the edge of Chat-Moss to Leigh, by which means new supplies of coal were introduced to Manchester from that district, and the traffic was still further increased. It was a saying of the Duke's, that "a navigation should always have coals at the heels of it."

The total cost of completing the canal from Worsley to Manchester, and from Longford Bridge to the Mersey at Runcorn, amounted to 220,000*l.* A truly magnificent undertaking, nobly planned and nobly executed. The power imparted by riches was probably never more munificently exercised than in this case; for, though the traffic proved a source of immense wealth to the Duke, it also conferred incalculable blessings upon the population of the district. It added much to their comforts, increased their employment, and facilitated the operations of industry in all ways. As soon as the canal was opened its advantages began to be felt. The charge for water-carriage between Liverpool and Manchester was lowered one-half. All sorts of produce were brought to the latter town, at moderate rates, from the farms and gardens adjacent to the navigation, whilst the value of agricultural property was immediately raised by the facilities afforded for the conveyance of lime and manure, as well as by reason of the more ready access to good markets which it provided for the farming classes. The Earl of Ellesmere has not less truly than elegantly observed, that "the history of Francis Duke of Bridgewater is engraved in intaglio on the face of the country he helped to civilize and enrich."

Probably the most remarkable circumstance connected with the money history of the enterprise is this: that although the canal yielded an income which eventually reached about 80,000*l.* a year, it was planned and executed by Brindley at a rate of pay considerably less than that of an ordinary mechanic of the present day. The highest wage he received whilst in the employment of the Duke was 3*s.* 6*d.* a day. For the greater part of the time he received only half-a-crown. Brindley, no doubt, accommodated himself to the Duke's pinched means, and the satisfactory completion of the canal was with him as much a matter of disinterested ambition and of professional character as of pay. He seems to have kept his own expenses down to the very lowest point. Whilst superintending the works at Longford Bridge, we find him making an entry of his day's personal charges at only 6*d.* for "ating and drink." On other days his outgoings were confined to "2*d.* for the turnpike." When living at "The Bull," near the works at Throstle Nest, we find his dinner costing 8*d.* and his breakfast 6*d.* His expenditure throughout was on an equally low scale, for he studied in all ways to economize the Duke's means, that every available shilling might be devoted to the prosecution of the works.

The Earl of Bridgewater, in his singular publication, the 'Letter to the Parisians,' above referred to, states that "Brindley offered to stay entirely with the Duke, and do business for no one else, if he would give him a guinea a week;" and this statement is repeated by the late Earl of Ellesmere in his 'Essays on History, Biography,' &c. But, on the face of it, the statement looks untrue; and we have since found, from Brindley's own note-book, that on the 25th of May, 1762, he was receiving a guinea a day from the Earl of Warrington for performing services for that nobleman; nor is it at all likely that he would prefer the Duke's three-and-sixpence a day to the more adequate rate of payment which he was accustomed to charge and to receive from other employers. It is quite true, however— and the fact is confirmed by Brindley's own record—that he received no more than a guinea a week whilst in the Duke's service; which only affords an illustration of the fact that eminent constructive genius may be displayed and engineering greatness achieved in the absence of any adequate material reward.

In a statement of the claims of Brindley's representatives, forwarded to the Earl of Bridgewater on the 3rd of November, 1803, it was stated that "during the period of his employ under His Grace, many highly advantageous and lucrative offers were made to him, particularly one from the Prince of Hesse, in 1766, who at that time was meditating a canal through his dominions in Germany, and who offered to subscribe to any terms Mr. Brindley might stipulate. To this engagement his family strongly urged him, but the solicitation of the Duke, in this as in every other instance, to remain with him, outweighed all pecuniary considerations; relying upon such a remuneration from His Grace as the profits of his work might afterwards justify." *

The inadequate character of his remuneration was doubtless well enough known to Brindley himself, and rendered him very independent in his bearing towards the Duke. They had frequent differences as to the proper mode of carrying on the works; but Brindley was quite as obstinate as the Duke on such occasions, and when he felt convinced

---

* We regret to have to add that Brindley's widow (afterwards the wife of Mr. Williamson, of Longport) in vain petitioned the Duke and his representatives, as well as the above Earl of Bridgewater, for payment of a balance said to have been due to Brindley for services, at the time of the engineer's death. In her letter to Robert Bradshaw, M.P., dated the 2nd May, 1803, Mrs. Williamson says : " It will doubtless appear to you extraordinary that so very late an application should now be made . . . but I must beg leave to state that repeated applications were made by me (after Mr. Brindley's sudden and unexpected death) to the late Mr. Thomas Gilbert and also to his brother, but without any other effect than that of constant promises to lay the matter before His Grace ; and I conceive it owing to this channel of application that no settling ever took place. A letter was also written to His Grace on this subject so late as the year 1801, out no answer was received. From the year 1765 to 1772, Mr. Brindley received no money on account of his salary. At that time he was frequently in very great want, and made application to the Duke, whose answer (to use the Duke's expression) was, ' I am much more distressed for money than you; however, as soon as I can recover myself, your services shall not go unrewarded.' In consequence of this, Mr. Brindley was under the necessity of borrowing several sums to make good engagements he was then under to various canal companies. In the year 1774, two years after Mr. Brindley's death, the late Mr. John Gilbert paid my brother, Mr. Henshall, the trifling sum of 100l. on account of Mr. Brindley's time, which is all that has been received. I beg leave to suggest how small and inadequate a return this is for his services during a period of seven years. Mr. B.'s travelling expenses on His Grace's account during that time were considerable, towards which, when he had not sufficient money to carry him the whole journey, he now and then received a small sum. How far his plans and undertakings have been beneficial to His Grace's interest is well known."

that his own plan was the right one he would not yield an inch. It is said that, after long evening discussions at the hearth of the old timbered hall at Worsley, or at the Duke's house at Liverpool, while the works there were in progress, the two would often part at night almost at daggers-drawn. But next morning, on meeting at breakfast, the Duke would very frankly say to his engineer, " Well, Brindley, I have been thinking over what we were talking about last night. I find you may be right after all; so just finish the work in your own way."

The Duke himself, to the end of his life, took the greatest personal interest in the working of his coal-mines, his canals, his mills, and his various branches of industry. These were his hobbies, and he took pleasure in nothing else. He was utterly lost to the fashionable world, and, as some thought, to a sense of its proprieties. Shortly after his canal had been opened for the conveyance of coals, the Duke established a service of passage-boats between Manchester and Worsley, and between Manchester and a station within two miles of Warrington, by which passengers were conveyed at the rate of a penny a mile.* The boats were fitted up like the Dutch treckschuyts, and, being found cheap as well as convenient, were largely patronized by the public. This service was afterwards extended to Runcorn, and from thence to Liverpool.

The Duke took particular pleasure in travelling by his own boats, preferring them to any more stately and aristo-cratic method. He often went by them to Manchester to watch how the coal-trade was going on. When the pas-sengers alighted at the coal-wharf, there were usually many poor people about, wheeling away their barrow-loads

---

* When the Duke had put on the boats and established the ser-vice, he offered to let them for 60l. a year; but not being able to find any person to take them at that price, he was under the necessity of conducting the service himself, by means of an agent. In the course of a short time the boats were found to yield a clear profit of 1600l. a year.

of coals. One of the Duke's regulations was, that whenever any deficiency in the supply was apprehended, those people who came with their wheelbarrows, baskets, and aprons for small quantities, should be served first, and waggons, carts, and horses sent away until the supply was more abundant. The numbers of small customers who thus resorted to the Duke's coal-yard rendered it a somewhat busy scene, and the Duke liked to look on and watch the proceedings.

One day a customer of the poorer sort, having got his sack filled, looked about for some one to help it on to his back. He observed a stoutish man standing near, dressed in a spencer, with dark drab smallclothes. "Heigh! mester!" said the man, "come, gie me a lift wi' this sack o' coal on to my shouder." Without any hesitation, the person in the spencer gave the man the required "lift," and off he trudged with the load. Some one near, who had witnessed the transaction, ran up to the man and asked, "Dun yo know who's that yo've been speaking tull?" "Naw! who is he?" "Why, it's th' Duke his-sen!" "The Duke!" exclaimed the man, dropping the bag of coals from his shoulder, "Hey! what'll he do *at* me? Maun a goo an ax his pardon?" But the Duke had disappeared.*

He was very fond of watching his men at work, especially when any new enterprise was on foot. When they were boring for coal at Worsley, the Duke came every morning and looked on for a long time together. The men did not like to leave off work whilst he remained there, and they became so dissatisfied at having to work so long beyond the hour at which the bell rang, that Brindley

---

* A similar story is told of him at Worsley. A boy had been fetching some coal from the mouth of the tunnel, and having rested his load could not get it cleverly on his back again. Seeing but not knowing the Duke, he called out, "Here, felly, gie us a hoist up." The Duke asked the boy a number of questions before helping him up with the sack; but when the youth felt it safe for carrying, he expressed his acknowledgments by observing "Wal, thou's a big chap, but thou's a lazy un!"

had difficulty in getting a sufficient number of hands to continue the boring. On inquiry, he found out the cause and communicated it to the Duke, who from that time made a point of immediately walking off when the bell rang, returning when the men had resumed work, and remaining with them usually until six o'clock. He observed, however, that though the men dropped work promptly as the bell rang, when he was not by, they were not nearly so punctual in resuming work, some straggling in many minutes after time. He asked to know the reason, and the men's excuse was, that though they could always hear the clock when it struck twelve, they could not so readily hear it when it struck only one. On this, the Duke had the mechanism of the clock altered so as to make it strike *thirteen* at one o'clock; which it continues to do to this day.

On another occasion, going into the yard at Worsley, he saw two men employed in grinding an axe, and three others looking on, probably waiting their turn at the grindstone. The Duke said nothing; but next morning he was in the yard early, and said to the foreman that he had observed it took five men to grind an axe. He then ordered that a water-wheel should be put up to drive the grindstone, and it was set about at once. The Duke was often after seen grinding the ferrule of his walking-stick against the self-acting machine.

His time was very fully occupied with his various business concerns, to which he gave a great deal of personal attention. Habit made him a business man—punctual in his appointments, precise in his arrangements, and economical both of money and time. When it was necessary for him to see any persons about matters of business, he preferred going to them instead of letting them come to him; "for," said he, "if they come to me, they may stay as long as they please; if I go to them, I stay as long as I please." His enforced habits of economy during the construction of the canal had fully impressed upon his mind the value of money. Yet, though "near," he was not penurious, but was usually liberal, and sometimes munificent. When the Loyalty Loan was raised, he contributed to it no less a sum than 100,000*l.* in cash. He was thoroughly and strongly national, and a generous patron of many public benevolent institutions.

The employer of a vast number of workpeople, he exercised his influence over them in such a manner as to evoke their gratitude and blessings. He did not "lord it" over them, but practically taught them, above all things, to help themselves. He was the pattern employer of his neighbourhood. With a kind concern for the welfare of his colliery workmen—then a half-savage class—he built comfortable dwellings and established shops and markets for them; by which he ensured that at least a certain portion of their weekly earnings should go to their wives and families in housing, food, and clothing, instead of being squandered in idle dissipation and drunkenness.

In order to put a stop to idle Mondays, he imposed a fine of half-a-crown on any workman who did not go down the pit at the usual hour on that morning; and hence the origin of what is called Half Crown Row at Worsley, as thus described by one of the colliers:—"T'ould dook fined ony mon as didn't go daown pit o' Moonday mornin auve a craown, and abeaut thot toime he made a new road to t'pit, so t'colliers caw'd it Auve Craown Row."

Debts contracted by the men at public-houses were not recognised by the pay-agents. The steadiest workmen were allowed to occupy the best and pleasantest houses as a reward for their good conduct. The Duke also bound the men to contribute so much of their weekly earnings to a general sick club; and he encouraged a religious tone of character amongst his people by the establishment of Sunday schools, which were directly superintended by his agents, selected from the best available class. The consequence was, that the Duke's colliers soon held a higher character for sobriety, intelligence, and good conduct, than the weavers and other workpeople of the adjacent country.

He did not often visit London, where he had long ceased to maintain a house; but when he went there he made an arrangement with one of his friends, who undertook for a stipulated sum to provide a daily dinner for His Grace and a certain number of guests whilst he remained in town. He also made occasional visits to his fine estate of Ashridge, in Buckinghamshire, taking the opportunity of spending a few days, going or coming, with Earl Gower and his Countess, the Duke's only sister, Lady Louisa Egerton, at Trentham Park. During his visits at the latter place, the Duke would get ensconced on a sofa in some distant corner of the room in the evenings, and discourse earnestly to those who would listen to him about the extraordinary advantages of canals. There was a good deal of fun made on these occasions about "the Duke's hobby." But he was always like a fish out of water until he got back to Worsley, to John Gilbert, his coal-pits, his drainage, his mills, and his canals.

No wonder he was fond of Worsley. It had been the scene of his triumphs, and the foundation of his greatness. Illustrious visitors from all parts resorted thither to witness Brindley's "castle in the air," and to explore the underground canals at Worsley-hill. Frisi, the Italian, the King of Denmark, and others, regarded these subterranean works with wonder and admiration when they were only from $1\frac{1}{2}$ to 2 miles in length; soon they extended to nearly 40 miles. Among the visitors entertained by the Duke was Fulton, the American artist, with whose speculations he was much interested. Fulton had given his attention to the subject of canals, and was then speculating on the employment of steam power for propelling canal boats. The Duke was so much impressed with Fulton's ingenuity, that he urged him to give up the profession of a painter and devote himself to that of a civil engineer. Fulton acted on his advice, and shortly after we find him residing at Birmingham—the central workshop of England —studying practical mechanics, and fitting himself for superintending the construction of canals, on which he was afterwards employed in the midland counties.*

The Duke did not forget the idea which Fulton had communicated to him as to the employment of steam as a motive power for boats, instead of horses; and when he afterwards heard that Symington's steam-boat, *The Dundas*, had been tried successfully on the Forth and Clyde Canal, he arranged to have six canal boats constructed after Symington's model; for he was a man to shrink from no expense in carrying out an enterprise which, to use his own words, had " utility at the heels of it." The Earl of Ellesmere, in his ' Essay on Aqueducts and Canals,' states that the Duke made actual experiment of a steam-tug, and quotes the following from the communication of one of the Duke's servants, alive in 1844: " I well remember the steam-tug experiment on the canal. It was between 1796 and 1799. Captain Shanks, R.N., from Deptford, was at Worsley many weeks preparing it, by the Duke's own orders and under his own eye. It was set going and tried with coal-boats; but it went slowly, and the paddles made sad work with the bottom of the canal, and also threw the water on the bank. The Worsley people called it Bonaparte."† But the Duke dying shortly after, the trustees refused to proceed with the experiment, and the project consequently fell through. Had the Duke lived, canal steam-tugs would doubtless have been fairly tried; and he might thus have initiated the practical introduction of steam-navigation in England, as he unquestionably laid the foundations of the canal system. He lived long enough, however, to witness the introduction of tram-roads, and he

---

* The treatise which Fulton afterwards published, entitled ' A Treatise on Canal Navigation, exhibiting the numerous advantages to be derived from small Canals, &c., with a description of the machinery for facilitating conveyance by water through the most mountainous countries, independent of Locks and Aqueducts,' (London, 1796,) is well known amongst engineers.

† Lord Ellesmere's ' Essays,' p. 241.

saw considerable grounds for apprehension in them. "We may do very well," he once observed to Lord Kenyon, "if we can keep clear of these —— tram-roads."

He was an admirable judge of character, and was rarely deceived as to the men he placed confidence in. John Gilbert was throughout his confidential adviser—a practical out-doors man, full of energy and perseverance. When any proposal was made to the Duke, he would say, "Well, thou must go to Gilbert and tell him all about it; I'll do nothing without I consult him." From living so much amongst his people, he had contracted their style of speaking, and "thee'd" and "thou'd" those whom he addressed, after the custom of the district. He was rough in his speech, and gruff and emphatic in his manner, like those amidst whom he lived; but with the rough word he meant and did the kindly act. His early want of education debarred him in a measure from the refining influences of letters; for he read little, except perhaps an occasional newspaper, and he avoided writing whenever he could. He also denied himself the graces of female society; and the seclusion which his early disappointment in love had first driven him to, at length grew into a habit. He lived wifeless and died childless. He would not even allow a woman servant to wait upon him.

In person he was large and corpulent; and the slim youth on whom the bet had been laid that he would be blown off his horse when riding the race in Trentham Park so many years before, had grown into a bulky and unwieldy man. His features strikingly resembled those of George III. and other members of the Royal Family. He dressed carelessly, and usually wore a suit of brown—something of the cut of Dr. Johnson's—with dark drab breeches, fastened at the knee with silver buckles. At dinner he rejected, with a kind of antipathy, all poultry, veal, and such like, calling them "white meats," and wondered that everybody, like himself, did not prefer the brown. He was a great smoker, and

Francis Duke of Bridgewater (æt. 70.)

smoked far more than he talked. Smoking was his principal evening's occupation when Brindley and Gilbert were pondering with him over the difficulty of raising funds to complete the navigation, and the Duke continued his solitary enjoyment through life. One of the droll habits to which he was addicted was that of rushing out of the room every five minutes, with the pipe in his mouth, to look at the barometer. Out of doors he snuffed, and he would pull huge pinches out of his right waistcoat pocket and thrust the powder up his nose, accompanying the operation with sundry strong short snorts.

He would have neither conservatory, pinery, flower-garden, nor shrubbery at Worsley; and once, on his return from London, finding some flowers which had been planted in his absence, he whipped their heads off with his cane, and ordered them to be rooted up. The only new things introduced about the place were some Turkey oaks, with which his character seemed to have more sympathy. But he took a sudden fancy for pictures, and with his almost boundless means the formation of a valuable collection of pictures was easy.*

Lord Ellesmere says: "An accident laid the foundation of the Bridgewater collection. Dining one day with his nephew, Lord Gower, afterwards Duke of Sutherland, the Duke saw and admired a picture which the latter had picked up a bargain, for some 10*l.*, at a broker's in the morning. 'You must take me,' he said, 'to that —— fellow to-morrow.' Whether this impetuosity produced any immediate result we are not informed, but plenty of such 'fellows' were doubtless not wanting to cater for the taste thus suddenly developed."

---

* " His purchases from Italy and Holland were judicious and important, and, finally, the distractions of France forcing the treasures of the Orleans Gallery into this country, he became a principal in the fortunate speculation of its purchase." — ' Essays on History, Biography,' &c.

Fortunately the Duke's investments in paintings appear to have been well directed; and a discerning eye seems to have guided a liberal hand in selecting fine separate works, as well as the gems from Continental collections which were then dispersed and found their way hither, thus enabling him to lay the foundation of the famous Bridgewater Gallery, one of the finest private collections in Europe. At his death, in 1803, its value was estimated at 150,000*l.*

The Duke very seldom took part in politics, but usually followed the lead of his relative Earl Gower, afterwards Marquis of Stafford, who was a Whig. In 1762, we find his name in a division on a motion to withdraw the British troops from Germany, and on the loss of the motion he joined in a protest on the subject. When the repeal of the American Stamp Act was under discussion His Grace was found in the ranks of the opposition to the measure. He strongly supported Mr. Fox's India Bill, and generally approved the policy of that statesman.

The title of Duke of Bridgewater died with him. The Earldom went to his cousin General Egerton, seventh Earl of Bridgewater, and from him to his brother the crazed Francis Henry, eighth Earl; and on his death at Paris, in February, 1829, that title too became extinct. The Duke bequeathed about 600,000*l.* in legacies to his relatives, General Egerton, the Countess of Carlisle, Lady Anne Vernon, and Lady Louisa Macdonald. He devised most of his houses, his pictures, and his canals, to his nephew George Granville (son of Earl Gower), second Marquis of Stafford and first Duke of Sutherland, with reversion to his second son, Lord Francis Egerton, first Earl of Ellesmere, who thus succeeded to the principal part of the vast property created by the Duke of Bridgewater. The Duke was buried in the family vault at Little Gaddesden, Hertfordshire, in the plainest manner, without any state, at his own express request. On his monument was inscribed the simple and appropriate epitaph *Impulit ille rates ubi duxit aratra Colonus.*

The Duke was a great public benefactor. The boldness of his enterprise, and the salutary results which flowed from its execution, entitle him to be regarded as one of the most useful men of his age. A Liverpool letter of 1765 says, "The services the Duke has rendered to the town and neighbourhood of Manchester have endeared him to the country, more especially to the poor, who, with grateful benedictions, repay their noble benefactor."[*] If he became rich through his enterprise, the public grew rich with him and by him; for his undertaking was no less productive to his neighbours than it was to himself. His memory was long venerated by the people amongst whom he lived,—a self-reliant, self-asserting race, proud of their independence, full of persevering energy, and strong in their attachments. The Duke was a man very much after their own hearts, and a good deal after their own manners. In respecting him, they were perhaps but paying homage to those qualities which they most cherished in themselves. Long after the Duke had gone from amongst them, they spoke to each other of his rough words and his kindly acts, his business zeal and his indomitable courage. He was the first great "Manchester man." His example deeply penetrated the Lancashire character, and his presence seems even yet to hover about the district. "The Duke's canal" still carries a large proportion of the merchandise of Manchester and the neighbouring towns; "the Duke's horses"[†] still draw "the

---

[*] 'History of Inland Navigation,' p. 76.

[†] The Duke at first employed mules in hauling the canal-boats, because of the greater endurance and freedom from disease of those animals, and also because they could eat almost any description of provender. The Duke's breed of mules was for a long time the finest that had been known in England. The popular impression in Manchester is, that the Duke's Acts of Parliament authorising the construction of his canals, forbade the use of horses, in order that men might be employed; and that the Duke consequently dodged the provisions of the Acts by employing mules. But this is not the case, there being no clause in any of them prohibiting the use of horses.

Duke's boats;" "the Duke's coals" still issue from "the Duke's levels;" and when any question affecting the traffic of the district is under consideration, the questions are still asked of "What will the Duke say?" "What will the Duke do?"*

Manchester men of this day may possibly be surprised to learn that they owe so much to a Duke, or that the old blood has helped the new so materially in the development of England's modern industry. But it is nevertheless true that the Duke of Bridgewater, more than any other single man, contributed to lay the foundations of the prosperity of Manchester, Liverpool, and the surrounding districts. The cutting of the canal from Worsley to Manchester conferred upon that town the immediate benefit of a cheap and abundant supply of coal; and when Watt's steam-engine became the great motive power in manufactures, such supply became absolutely essential to its existence as a manufacturing town. Being the first to secure this great advantage, Manchester thus got the start forward which she has never since lost.†

But, besides being a waterway for coal, the Duke's canal, when opened out to Liverpool, immediately conferred upon Manchester the immense advantage of direct connection with an excellent seaport. New canals, supported by the Duke and constructed by the Duke's engineer, grew out of the original scheme between Manchester and Runcorn, which had the further effect of placing the former town in direct water-communication with the rich districts of the north-west of England. Then the Duke's

---

* There is even a tradition surviving at Worsley, that "the Duke" rides through the village once in every year at midnight, drawn by six coal-black horses!

† The cotton trade was not of much importance at first, though it rapidly increased when the steam-engine and spinning-jenny had become generally adopted. It may be interesting to know that sixty years since it was considered satisfactory if *one* cotton-flat a day reached Manchester from Liverpool. In the Duke's time the flats always "cast anchor" on their way, or at least laid up for the night, at six o'clock precisely, starting again at six o'clock on the following morning.

canal terminus became so important, that most of the new navigations were laid out to join it; those of Leigh, Bolton, Stockport, Rochdale, and the West Riding of Yorkshire, being all connected with the Duke's system, whose centre was at Manchester. And thus the whole industry of these districts was brought, as it were, to the very doors of that town.

But Liverpool was not less directly benefited by the Duke's enterprise. Before his canal was constructed, the small quantity of Manchester woollens and cottons manufactured for exportation was carried on horses' backs to Bewdley and Bridgenorth on the Severn, from whence they were floated down that river to Bristol, then the chief seaport on the west coast. No sooner, however, was the new water-way opened out than the Bridgenorth pack-horses were taken off, and the whole export trade of the district was concentrated at Liverpool. The additional accommodation required for the increased business of the port was promptly provided as occasion required. New harbours and docks were built, and before many years had passed Liverpool had shot far ahead of Bristol, and became the chief port on the west coast, if not in all England. Had Bristol been blessed with a Duke of Bridgewater, the result might have been altogether different; and the valleys of Wilts, the coal and iron fields of Wales, and the estuary of the Severn, might have been what South Lancashire and the Mersey are now. Were statues any proof of merit, the Duke would long since have had the highest statue in Manchester as well as Liverpool erected to his memory, and that of Brindley would have been found standing by his side; for they were both heroes of industry and of peace, though even in commercial towns men of war are sometimes more honoured.

We can only briefly glance at the extraordinary growth of Manchester since the formation of the Duke's canal, as indicated by the annexed plan.

Though Manchester was a place of some importance about the middle of last century, it was altogether insignificant in extent, trade, and population, compared with

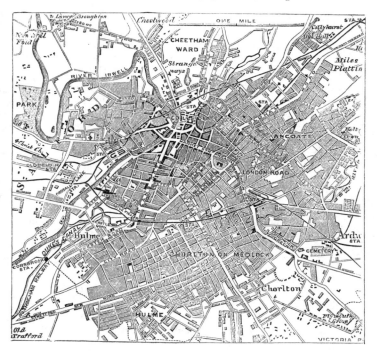

Plan of Manchester, showing its extent at three periods.

(The parts printed black ▐ represent Manchester in 1770; those dark-shaded ▐ show its

extent in 1804; and the light-shaded parts ▦ Manchester at the present day.)

what it is now. It consisted of a few principal streets—narrow, dark, and tortuous—one of them leading from the Market Place to St. Ann's Square, being very appropriately named "Dark Entry." Deansgate was the principal original street of the town, and so called because of its leading to the dean or valley along which it partly extended. From thence a few streets diverged in different directions into the open country. St. Ann's Square, the fashionable centre of modern Manchester, was in 1770 a corn-field surrounded with lofty trees, and known by the name of "Acre's Field." The cattle-fairs of the town were held there, the entrance from Deansgate being by Toll Lane, a narrow, dirty, unpaved way, so called because toll was there levied on the cattle proceeding towards the fair. The ancient seat of the Radcliffe family still stood at Pool Fold, close to the site of the modern Cross Street, and the water in the moat was used as a ducking-pond for scolds. When the pool became filled up, the ducking-pond was removed to Daub Holes, then on the outskirts of the town, where the Infirmary now stands. The site of King Street, now the very heart of Manchester, was as yet comparatively retired, a colony of rooks having established themselves in the tall trees at its upper end, from which they were only finally expelled about forty years ago. Cannon Street was the principal place of business, the merchants and their families living in the comparatively humble tenements fronting the street, the equally humble warehouses in which their business was done standing in the rear. The ground on which the crowded thoroughfares of Oldham Street, London Road, Mosley Street, and their continuations, now exist, was as yet but garden or pasture-land. Salford itself was only a hamlet occupying the bend of the Irwell. It consisted of a double line of mean houses, extending from the Old Bridge (now Victoria Bridge) to about the end of Gravel Lane, then a country road containing only a few detached cottages. The comparatively rural character of Manchester may be inferred from the circumstance that the Medlock and the Irk, the Tib and Shooter's Brook, were favourite fishing streams. Salmon were caught in the Medlock and at the mouth of the Irk; and the others were well stocked with trout.

The Medlock and the Irk are now as black as old ink, and as thick; but the Tib and Shooter's Brook are entirely lost,—having been absorbed, like the London Fleet, in the sewage system of the town. Tib Street and Tib Lane indicate the former course of the Tib; but of Shooter's Brook not a trace is left.

The townships of Ardwick Green, Hulme, and Chorlton-upon-Medlock (formerly called Chorlton Row), were entirely rural. The old rate-books of Chorlton Row exhibit some curious facts as to the transformations effected in that township. In 1720, a "lay" of 14d. in the pound produced a sum of 26l. 18s., the whole disbursements for the year amounting to 28l. 8s. 5d. From the highway rate laid in 1722, it appears that the contributors were only twenty persons in all, whose payments ranged from 8d. to 1l. 13s. 4d., producing a total levy of 6l. 18s. 10d. for the year. From the disbursements, it appears that the regular wage paid to the workmen employed was a shilling a-day. In 1750, a lay of 3d. in the pound produced only 6l. 2s. 1½d.; so that the population and value of property in Chorlton Row had not much increased during the thirty years that had passed. In 1770, two levies brought in 57l. 8s. 6d.; and in 1794, four, made in that year, produced 208l. 2s. 4d.* Among the list of contributors in the latter year we find " Mrs. Quincey 16s. 6d." —the mother of De Quincey, the English opium-eater, who was brought up in Chorlton Row. De Quincey describes the home of his childhood as a solitary house, "beyond which was nothing but a cluster of cottages, composing the little hamlet of Greenhill." It was connected by a winding lane with the Rusholme road. The house, called Greenheys—the nucleus of an immense suburban district—built by De Quincey's father, "was then,"

---

* Recent "poor-lays" exhibit a very different result from what they lid in former years. In 1860–1 the poor-rate levied in Chorlton-upon-Medlock yielded (at 2s. 10d. in the pound) 18,798l.; the property in the township being of the rateable value of 145,844l.

he says, "a clear mile from the outskirts of Manchester," Princess Street being then the termination of the town on that side.* Now it is enveloped by buildings in all directions, and nothing of the former rural character of the neighbourhood remains but the names of Greenhill, Rusholme, and Greenheys.

Coming down to the second expansion of Manchester, as exhibited on our plan, it will be observed that a considerable increase of buildings had taken place in the interval between 1770 and 1804. The greater part of the town was then contained in the area bounded by Deansgate, the crooked lanes leading to Princess Street, Bond Street, and David Street, to the Rochdale Canal, and round by Ancoats Lane (now Great Ancoats Street) and Swan Street, to Long Millgate, then a steep narrow lane forming the great highway into North Lancashire. Very few buildings existed outside the irregular quadrangle indicated by the streets we have named. The straggling houses of Deansgate, which were principally of timber, ended at Knott Mill. A few dye-works stood at intervals along the Medlock, now densely occupied by buildings for miles along both banks. Salford had not yet extended to St. Stephen's Street in one direction, nor above half way to Broughton Bridge in another.† The comparatively limited spaces thus indicated sufficed, however, for places of business and habitations for the population. Now the central districts are almost exclusively occupied for business purposes, and houses for dwellings have rapidly extended in all directions. The populous districts of Broughton, Higher and Lower, did not exist thirty-five years ago. They contained no buildings excepting Strange-

---

* De Quincey's 'Autobiographic Sketches,' pp. 34, 48.

† The corner of Irwell Street, Salford, as recently as 1828, was occupied by an old canal "flat," tenanted by an eccentric character, after whom it was designated "Bannister's Ship." Opposite it was a row of cottages with gardens in front. Oldfield and Ordsall Lanes were country roads, and the streets adjacent to them were not yet in existence.

ways Hall and a few cottages which lay scattered beyond the bottom of the workhouse brow; the locality where the new Assize Courts have been erected, which the citizens of Manchester claim to be unequalled in the kingdom for magnificence and accommodation.

But pastures, corn-fields, and gardens rapidly gave place to streets and factory buildings.* The suburban districts of Ardwick, Hulme, and Cheetham, became wholly absorbed in the great city. Stretford New Road, a broad street nearly a mile and a half long, forms the main highway for a district occupied during the life of the present generation by a population greater than that of many cities. Not fifty years since, a few farm-houses and detached dwellings were all the buildings it contained, and Chester Road, the principal one in the district, was a narrow winding lane, with hedges on each side. Jackson's Lane, remembered as a mere farm-road through corn-fields, has become a spacious thoroughfare dignified with the name of Great Jackson Street, that contains a relic of rural Hulme in the remnant of "Jackson's Farm" buildings, which gave the name, first to the "lane" and then to the "street." It is a single-storey building, covered with grey flags, and stands in an oblique recess on the left-hand side, about half-way between Chester Road and the recently formed City Road. Higher up, at the junction of Chapman and Preston Streets, the houses, also covered with grey flags, still remain, which, within a comparatively recent period, stood amidst fields, and were known as "Geary's Farm,"—these buildings are now surrounded by streets on every side. About thirty years since, the part of Hulme nearest to Manchester was occupied by "tea gardens," and places of resort much used by the "town" population. The

---

* The growth of Manchester, and the sister borough of Salford, will be more readily appreciated, perhaps, by a glance at the population at different periods than by any other illustration:—

Population in	1774.	1801.	1821.	1861.
	41,032	84,020	187,031	460,028.

principal of these was at the White House; and it is said of late roysterers at that place, that unless they could form a party or secure the services of " the patrol," they had frequently to sojourn there all night. The officers constituting the patrol * carried swords and horn lanterns; and, clad as they were in heavy greatcoats with many capes, they were by no means light of foot, or at all formidable adversaries to the footpads who " worked " the district.

Among the most remarkable improvements in Manchester of late years, have been the numerous spacious thoroughfares which have been opened up in all directions. In this respect, the public spirit of Manchester has not been surpassed by any town in the kingdom,— the new streets being laid out on a settled plan with a view to future extension, and executed with admirable judgment. Narrow, dark, and crooked ways have been converted into wide and straight streets, admitting light, air, and health to the inhabitants, and affording spacious highways for the great and growing traffic of the district. The important street-improvements executed in Manchester during the last thirty years have cost an aggregate of about 800,000l. The central and oldest part of the town has thus undergone a complete transformation. So numerous are the dark and narrow entries that have been opened up—the obstructive buildings that have been swept away, the projecting angles that have been cut off, and the crooked ways that have been made straight—that the denizen of a former age would be very unlikely to recognise the Manchester of to-day, were it possible for him to revisit it.

Some of the street-improvements have their peculiar social aspects, and call up curious reminiscences. The

---

* On the subject of watchmen, it may be mentioned that the first watchman was appointed for Chorlton-on-Medlock in 1814. In 1832 an Act was obtained for improving and regulating that township, and so recently as 1833 it was first lighted with gas. The Police Act for the Township of Hulme was obtained in 1834.

stocks, pillory, and Old Market Cross, were removed from
the Market Place in 1816. The public whipping of cul-
prits on the pillory stage is within the recollection of the
elder portion of the present inhabitants. Another " social
institution," of a somewhat different character, was ex-
tinguished much more recently, by the construction of
the splendid piece of terrace-road in front of the cathedral, known as the Hunt's Bank improvement. This
road swept away a number of buildings, shown on the
old plans of Manchester as standing on the water's edge,
close to the confluence of the Irk with the Irwell. They
were reached by a flight of some thirty steps, and consisted of a dye-work, employing three or four hands, two
public-houses, and about a dozen cottages and other buildings. The public-houses, the 'Ring o' Bells' and the
' Blackamoor,' particularly the former, were famous places
in their day. On Mondays, wedding-parties from the
country, consisting sometimes of from twenty to thirty
couples, accompanied by fiddlers, visited " t' Owd Church"
to get married. The 'Ring o' Bells' was the rendez-
vous until the parties were duly married and ready to
form and depart homewards, in a more or less orderly
manner, headed by their fiddlers as they had come. The
'Ring o' Bells' was also a favourite resort of the re-
cruiting-serjeant, and more recruits, it is said, were en-
listed there than at any other public-house in the kingdom. But these, and many curious characteristics of old
Manchester, have long since passed away; and not only
the town but its population have become entirely new.

Bridgewater Halfpenny Token.

CHAPTER VI.

BRINDLEY CONSTRUCTS THE GRAND TRUNK CANAL.

LONG before the Duke's Canal was finished, Brindley was actively employed in carrying out a still larger enterprise,—a canal to connect the Mersey with the Trent, and both with the Severn,—thus uniting by a grand line of water-communication the ports of Liverpool, Hull, and Bristol. He had, as we have already seen, made a survey of such a canal, at the instance of Earl Gower, before his engagement as engineer for the Bridgewater undertaking. Thus, in the beginning of February, 1758, before the Duke's bill had been even applied for, we find him occupied for days together "a bout the novogation," and he then surveyed the country between Longbridge in Staffordshire, and King's Mills in Derbyshire.

The enterprise, however, made very little progress. The success of canals in England was as yet entirely problematical; and this was of too formidable a character to be hastily undertaken. But again, in 1759, we find Brindley proceeding with his survey of the Staffordshire Canal; and in the middle of the following year he was occupied about twenty days in levelling from Harecastle, at the summit of the proposed canal, to Wilden, near Derby. During that time he had many interviews with Earl Gower at Trentham, and with the Earl of Stamford at Enville, discussing the project.

The next step taken was the holding of a public meeting at Sandon, in Staffordshire, as to the proper course which the canal should take, if finally decided upon. Considerable difference of opinion was expressed at the meeting, in consequence of which it was arranged that Mr. Smeaton should be called upon to co-operate with Brindley in making a joint survey and a joint report.

A second meeting was held at Wolseley Bridge, at which the plans of the two engineers were ordered to be engraved and circulated amongst the landowners and others interested in the project. Here the matter rested for several years more, without any further action being taken. Brindley was hard at work upon the Duke's Canal, and the Staffordshire projectors were disposed to wait the issue of that experiment; but no sooner had it been opened, and its extraordinary success become matter of fact, than the project of the canal through Staffordshire was again revived. The gentlemen of Cheshire and Staffordshire, especially the salt manufacturers of the former county and the earthenware-manufacturers of the latter, now determined to enter into co-operation with the leading landowners in concerting the necessary measures with the object of opening up a line of water-communication with the Mersey and the Trent.

The earthenware manufacture, though in its infancy, had already made considerable progress; but, like every other branch of industry in England at that time, its further development was greatly hampered by the wretched state of the roads. Throughout Staffordshire they were as yet, for the most part, narrow, deep, circuitous, miry, and inconvenient; barely passable with rude waggons in summer, and almost impassable, even with pack-horses, in winter. Yet the principal materials used in the manufacture of pottery, especially of the best kinds, were necessarily brought from a great distance — flint-stones from the south-eastern ports of England, and clay from Devonshire and Cornwall. The flints were brought by sea to Hull, and the clay to Liverpool. From Hull the materials were brought up the Trent in boats to Willington; and the clay was in like manner brought from Liverpool up the Weaver to Winsford, in Cheshire. Considerable quantities of clay were also conveyed in boats from Bristol, up the Severn, to Bridgenorth and Bewdley. From these various points the materials were conveyed by land-carriage, mostly on the backs of horses, to the towns in the Potteries, where they were worked up into earthenware and china.

The manufactured articles were returned for export in the same rude way. Large crates of pot-ware were slung across horses' backs, and thus conveyed to their respective ports, not only at great risk of breakage and pilferage, but also at a heavy cost. The expense of carriage was not less than a shilling a ton per mile, and the lowest charge was eight shillings the ton for ten miles. Besides, the navigation of the rivers above mentioned was most uncertain, arising from floods in winter and droughts in summer. The effect was, to prevent the expansion of the earthenware manufacture, and very greatly to restrict the distribution of the lower-priced articles in common use.

The same difficulty and cost of transport checked the growth of nearly all other branches of industry, and made living both dear and uncomfortable. The indispensable article of salt, manufactured at the Cheshire Wiches, was in like manner carried on horses' backs all over the country, and reached almost a fabulous price by the time it was sold two or three counties off. About a hundred and fifty pack-horses, in gangs, were occupied in going weekly from Manchester, through Stafford, to Bewdley and Bridgenorth, loaded with woollen and cotton cloth for exportation ;* but the cost of the carriage by this

---

* In a curious book published in 1766, by Richard Whitworth, of Balcham Grange, Staffordshire, afterwards Sir Richard Whitworth, member for Stafford, entitled 'The Advantages of Inland Navigation,' he points to the various kinds of traffic that might be expected to come upon the canal then proposed by him, and amongst other items he enumerates the following :—

"There are three pot-waggons go from Newcastle and Burslem weekly, through Eccleshall and Newport to Bridgenorth, and carry about eight tons of pot-ware every week, at 3l. per ton. The same waggons load back with ten tons of close goods, consisting of white clay, grocery, and iron, at the same price, delivered on their road to Newcastle. Large quantities of pot-

mode so enhanced the price, that it is clear that in the
case of many articles it must have acted as a prohibition,
and greatly checked both production and consumption.
Even corn, coal, lime, and iron-stone were conveyed in
the same way, and the operations of agriculture, as of
manufacture, were alike injuriously impeded.   There were
no shops then in the Potteries, the people being supplied
with wares and drapery by packmen and hucksters, or
from Newcastle-under-Lyne, which was the only town in
the neighbourhood worthy of the name.

The people of the district in question were quite as
rough as their roads.   Their manners were coarse, and
their amusements brutal.   Bull-baiting, cock-throwing, and
goose-riding were their favourite sports.   When Wesley
first visited Burslem, in 1760, the potters assembled to jeer
and laugh at him.   They then proceeded to pelt him.   "One
of them," he says, "threw a clod of earth which struck
me on the side of the head; but it neither disturbed me
nor the congregation."   At that time the whole population
of the Potteries did not amount to more than about
7000 souls.   The villages in which they lived were poor
and mean, scattered up and down, and the houses were
mostly covered with thatch.   Hence the Rev. Mr.
Middleton, incumbent of Stone—a man of great shrewd-
ness and quaintness, distinguished for his love of harmless
mirth and sarcastic humour—when enforcing the duty of
humility upon his leading parishioners, took the opportunity, on one occasion, after the period of which we
speak, of reminding them of the indigence and obscurity

---

ware are conveyed on horses' backs
from Burslem and Newcastle to
Bridgenorth and Bewdley for exportation—about one hundred tons
yearly, at 2l. 10s. per ton.   Two
broad-wheel waggons (exclusive of
150 pack-horses) go from Manchester through Stafford weekly,
and may be computed to carry

312 tons of cloth and Manchester
wares in the year, at 3l. 10s. per
ton.   The great salt-trade that is
carried on at Northwich may be
computed to send 600 tons yearly
along this canal, together with
Nantwich 400, chiefly carried now
on horses' backs, at 10s. per ton
on a medium."

from which they had risen to opulence and respectability.
He said they might be compared to so many sparrows,
for that all of them had been hatched *under the thatch.*
When the congregation of this gentleman, growing rich,
bought an organ and placed it in the church, he persisted
in calling it the hurdy-gurdy, and often took occasion to
lament the loss of his old psalm-singers.

The people towards the north were no better, nor were
those further south. When Wesley preached at Congleton,
four years later, he said, " even the poor potters [though
they had pelted him] are a more civilized people than the
better sort (so called) at Congleton." Arthur Young
visited the neighbourhood of Newcastle-under-Lyne in
1770, and found poor-rates high, wages low, and employment scarce. " Idleness," said he, " is the chief employment of the women and children. All drink tea, and fly
to the parishes for relief at the very time that even a
woman for washing is not to be had. By many accounts
I received of the poor in this neighbourhood, I apprehend
the rates are burthened for the spreading of laziness,
drunkenness, tea-drinking, and debauchery,—the general
effect of them, indeed, all over the kingdom." *

Hutton's account of the population inhabiting the
southern portion of the same county is even more dismal.
Between Hales Owen and Stourbridge was a district
usually called the Lie Waste, and sometimes the Mud
City. Houses stood about in every direction, composed
of clay scooped out into a tenement, hardened by the sun,
and often destroyed by the frost. The males were half-
naked, the children dirty and hung over with rags. " One
might as well look for the moon in a coal-pit," says
Hutton, " as for stays or white linen in the City of Mud.
The principal tool in business is the hammer, and the
beast of burden the ass." †

---

* Young's 'Six Months' Tour.'   † 'History of Birmingham.' Ed.
Ed. 1770. Vol. iii., p. 317.   1836, p. 24.

The district, however, was not without its sprinkling of public-spirited men who were actively engaged in devising new sources of employment for the population; and, as one of the most effective means of accomplishing this object, opening up the communications, by road and canal, with near as well as distant parts of the country. One of the most zealous of such workers was the illustrious Josiah Wedgwood. He was one of those indefatigable men who from time to time spring from the ranks of the common people, and by their energy, skill, and enterprise, not only practically educate the working population in habits of industry, but, by the example of diligence and perseverance which they set before them, largely influence public action in all directions, and contribute in a great measure to form the national character.

Josiah Wedgwood was born in a humble position in life; and though he rose to eminence as a man of science as well as a manufacturer, he possessed no greater advantages at starting than Brindley himself did. His grandfather and granduncle were both potters, as was also his father Thomas, who died when Josiah was a mere boy, the youngest of a family of thirteen children. He began his industrial life as a thrower in a small pot-work, conducted by his elder brother; and he might have continued working at the wheel but for an attack of virulent small-pox, which, being neglected, led to a disease in his right leg, which in a great measure unfitted him for following even that humble employment. When he returned to his work, most probably before he was sufficiently recovered from his illness, the pain in his limb was such that he had to keep it almost constantly rested upon a stool before him.* The disease continued increasing as he advanced

---

* The Papers of Mr. Llewellynn Jewitt, F.S.A., on ' Wedgwood and Etruria,' published in the *Art Journal*, 1864, contain many in- teresting particulars relating to the life and labours of Wedgwood, and are well worthy of perusal.

in years, and it was greatly aggravated by an unfortunate bruise, which kept him to his bed for months, and reduced him to the last extremity of debility. At length the disorder reached the knee, and threatened to endanger his life, when amputation was found necessary. During the enforced leisure of his many illnesses arising from this cause, Wedgwood took to reading and thinking, and turned over in his mind the various ways of making a living by his trade, now that he could no longer work at the potter's wheel. It has been no less elegantly than truthfully observed by Mr. Gladstone, that "it is not often that we have such palpable occasion to record our obligations to the small-pox. But in the wonderful ways of Providence, that disease, which came to him as a twofold scourge, was probably the occasion of his subsequent excellence. It prevented him from growing up to be the active, vigorous English workman, possessed of all his limbs, and knowing right well the use of them; but it put him upon considering whether, as he could not be that, he might not be something else, and something greater. It sent his mind inwards; it drove him to meditate upon the laws and secrets of his art. The result was, that he arrived at a perception and a grasp of them which might, perhaps, have been envied, certainly have been owned, by an Athenian potter." *

Wedgwood began operations on his own account by making various ornamental objects out of potter's clay, such as knife-hafts, boxes, and sundry curious little articles for domestic use. He joined in several successive partnerships with other workmen, but made comparatively small progress until he began business for himself, in 1759, in a humble cottage near the Market House in Burslem, known by the name of the Ivy House. He

---

* 'Wedgwood;' an Address delivered at Burslem, Oct. 26, 1863. By the Right Hon. W. E. Glad-stone, Chancellor of the Exchequer. London: Murray.

there pursued his manufacture of knife-handles and other small wares, striving at the same time to acquire such a knowledge of practical chemistry as might enable him to

Ivy House, Burslem, Wedgwood's first Pottery.*

improve the quality of his work in respect of colour, glaze, and durability. Success attended Wedgwood's diligent and persistent efforts, and he proceeded from one stage of improvement to another, until at length, after a course of about thirty years' labour, he firmly established a new branch of industry, which not only added greatly to the conveniences of domestic life, but proved a source of remunerative employment to many thousand families throughout England.

---

* The Ivy House, in which Wedgwood began business on his own account, is the cottage shown on the right-hand of the engraving. The other house is the old "Turk's Head."

His trade having begun to expand, an extensive demand for his articles sprang up, not only in London, but in foreign countries.* But there was this great difficulty in his way,—tnat the roads in his neighbourhood were so bad that he was at the same time prevented from obtaining a sufficient supply of the best kinds of clay and also from disposing of his wares in distant markets. This great evil weighed heavily upon the whole industry of the district, and Wedgwood accordingly appears to have bestirred himself at an early period in his career to improve the local communications.   In conjunction with several of the leading potters he promoted an application to Parliament for powers to repair and widen the road from the Red Bull, at Lawton, in Cheshire, to Cliff Bank, in Staffordshire. This line, if formed, would run right through the centre of the Potteries, open them to traffic, and fall at either end into a turnpike road.

The measure was, however, violently opposed by the people of Newcastle-under-Lyne, on the ground that the proposed new road would enable waggons and packhorses to travel north and south from the Potteries without passing through their town. The Newcastle innkeepers acted as if they had a vested interest in the bad roads; but the bill passed, and the new line was made, stopping short at Burslem. This was, no doubt, a great advantage, but it was not enough. The heavy carriage of clay, coal, and earthenware needed some more convenient means of

---

* Faujas Saint Fond, in his 'Travels in England,' thus writes respecting Wedgwood's ware :—" Its excellent workmanship, its solidity, the advantage which it possesses of standing the action of fire, its fine glaze, impenetrable to acids, the beauty, convenience, and variety of its forms, and its moderate price, have created a commerce so active, and so universal, that in travelling from Paris to St. Petersburg, from Amsterdam to the farthest point of Sweden, from Dunkirk to the southern extremity of France, one is served at every inn from English earthenware. The same fine article adorns the tables of Spain, Portugal, and Italy; and it provides the cargoes of ships to the East Indies, the West Indies, and America."

Josiah Wedgwood.

transport than waggons and roads; and, when the subject of water communication came to be discussed, Josiah Wedgwood at once saw that a canal was the very thing for the Potteries. Hence he immediately entered with great spirit into the movement again set on foot for the construction of Brindley's Grand Trunk Canal.

The field was not, however, so clear now as it had been before. The success of the Duke's canal led to the projection of a host of competing schemes in the county of

Chester, and it appeared that Brindley's Grand Trunk project would have to run the gauntlet of a powerful local opposition. There were two other schemes besides his, which formed the subject of much pamphleteering and controversy at the time, one entering the district by the river Weaver, and another by the Dee. Neither of these proposed to join the Duke of Bridgewater's canal, whereas the Grand Trunk line was laid out so as to run into his at Preston-on-the-Hill near Runcorn. As the Duke was desirous of placing his navigation—and through it Manchester, Liverpool, and the intervening districts—in connection with the Cheshire Wiches and the Staffordshire Potteries, he at once threw the whole weight of his support upon the side of Brindley's Grand Trunk. Indeed, he had himself been partly at the expense of its preliminary survey, as we find from an entry in Brindley's memorandum-book, under date the 12th of April, 1762, as follows: " Worsley—Rec^d from Mr Tho Gilbert for ye Staffordshire survey, on account, 33*l*. 16*s*. 11*d*."

The Cheshire gentlemen protested against the Grand Trunk scheme, as calculated to place a monopoly of the Staffordshire and Cheshire traffic in the hands of the Duke; but they concealed the fact, that the adoption of their respective measures would have established a similar monopoly in the hands of the Weaver Canal Company, whose line of navigation, so far as it went, was tedious, irregular, and expensive. Both parties mustered their forces for a Parliamentary struggle, and Brindley exerted himself at Manchester and Liverpool in obtaining support and evidence on behalf of his plan. The following letter from him to Gilbert, then at Worsley, relates to the rival schemes.

" 21 Decr. 1765

" On Tusdey Sr Georg [Warren] sent Nuton in to Manchester to make what intrest he could for Sir Georg and to gather ye old Navogtors togather to meet Sir Georg at Stoperd to make Head a ganst His Grace

"I sawe Docter Seswige who sese Hee wants to see you about pamant of His Land in Cheshire

"On Wednesday ther was not much transpired but was so dark I could carse do aneything

"On Thursdey Wadgwood of Burslam came to Dunham & sant for mee and wee dined with Lord Gree [Grey] & Sir Hare Mainwering and others   Sir Hare cud not ceep His Tamer [temper]   Mr. Wedgwood came to seliset Lord Gree in faver of the Staffordshire Canal & stade at Mrs Latoune all night & I whith him & on frydey sat out to wate on Mr Edgerton to seliset Him   Hee sase Sparrow and others are indavering to gat ye Land owners consants from Hare Castle to Agden

"I have ordered Simcock to ye Langth falls of Sanke Navegacion.

"Ryle wants to have coals sant faster to Alteringham that Hee may have an opertunety dray of ye sale Moor Canal in a bout a weeks time.

"I in tend being back on Tusdy at fardest."

The first public movement was made by the supporters of Brindley's scheme. They held an open meeting at Wolseley Bridge, Staffordshire, on the 30th of December, 1765, at which the subject was fully discussed. Earl Gower, the lord-lieutenant of the county, occupied the chair; and Lord Grey and Mr. Bagot, members for the county,—Mr. Anson, member for Lichfield,—Mr. Thomas Gilbert, the agent for Earl Gower, then member for Newcastle-under-Lyne,—Mr. Wedgwood, and many other influential gentlemen, were present to take part in the proceedings. Mr. Brindley was called upon to explain his plans, which he did to the satisfaction of the meeting; and these having been adopted, with a few immaterial alterations, it was determined that steps should be taken to apply for a bill conferring the necessary powers in the next session of Parliament. Mr. Wedgwood put his name down for a thousand pounds towards the preliminary expenses, and promised to subscribe largely for shares besides.* The promoters of the measure proposed to designate the undertaking " The Canal from the Trent to the Mersey ;" but Brindley, with sagacious foresight, urged that it should be called The Grand Trunk, because, in his judgment, numerous other canals would branch out from it at various points of its course, in like manner as the arteries of the human system branch out from the aorta ; and before many years had passed, his anticipations in this respect were fully realized. The Staffordshire potters were greatly pleased with the decision of the meeting, and on the following evening they assembled round a large bonfire at Burslem, and drank the healths of Lord Gower, Mr. Gilbert, and the other promoters of the scheme, with fervent demonstrations of joy.

The opponents of the measure also held meetings, at which they strongly declaimed against the Duke's proposed monopoly, and set forth the superior merits of their respective schemes. One of these was a canal from the river Weaver, by Nantwich, Eccleshall, and Stafford, to the Trent at Wilden Ferry, without touching the Potteries at all. Another was for a canal from the Weaver at Northwich, passing by Macclesfield and Stockport, round to Manchester, thus completely surrounding the Duke's navigation, and preventing its extension southward into Staffordshire or any other part of the Midland districts.

But there was also a strong party opposed to all canals

---

* Wedgwood even entered the lists as a pamphleteer in aid of the Grand Trunk project, and, in 1765, he and his partner, Mr. Bentley, formerly of Liverpool, drew up a very able statement, showing the advantages likely to be derived from the construction of the proposed canal, under the title of ' A View of the Advantages of Inland Navigation, with a plan of a Navigable Canal intended for a communication between the ports of Liverpool and Hull.' It pointed out in glowing language the advantages to be derived from opening up the internal communications of a country by means of roads, canals, &c.; and showed how the comfort and even the necessity of all classes must be so much better provided for by a reduction in the cost of carriage of useful and necessary commodities.

whatever—the party of croakers, who are always found in opposition to improved communications, whether in the shape of turnpike roads, canals, or railways. They prophecied that if the proposed canals were made, the country would be ruined, the breed of English horses would be destroyed, the innkeepers would be made bankrupts, and the pack-horses and their drivers would be deprived of their subsistence. It was even said that the canals, by putting a stop to the coasting trade, would destroy the race of seamen. It is a fortunate thing for England that it has contrived to survive these repeated prophecies of ruin. But the manner in which our countrymen contrive to grumble their way along the high road of enterprise, thriving and grumbling, is one of the peculiar features in our character which perhaps only Englishmen can understand and appreciate.

It is a curious illustration of the timidity with which the projectors of those days entered upon canal enterprise, that one of their most able advocates, in order to mitigate the opposition of the pack-horse and waggon interest, proposed that "no main trunk of a canal should be carried nearer than within four miles of any great manufacturing and trading town; which distance from the canal would be sufficient to maintain the same number of carriers and to employ almost the same number of horses as before." * But as none of the towns in the Potteries were as yet large manufacturing or trading places, this objection did not apply to them, nor prevent the canals from being carried quite through the centre of what has since become a continuous district of populous manufacturing towns and villages. The vested interests of some of the larger towns were, however, for this reason, preserved, greatly to their own ultimate injury; and when the canal, to conciliate the local opposition, was so laid out as to leave them at a distance, not many

---

* 'The Advantages of Inland Navigation,' by R. Whitworth. 1766.

years elapsed before they became clamorous for branches
to join the main trunk — but not until the mischief had
been done, and a blow dealt to their own trade, in consequence of their being left so far outside the main line
of water communication, from which many of them never
after recovered.

It is not necessary to describe the Parliamentary contest upon the Grand Trunk Canal Bill.   There was the
usual muster of hostile interests,—the river navigation
companies uniting to oppose the new and rival company—
the array of witnesses on both sides, — Brindley, Wedgwood, Gilbert, and many more, giving their evidence in
support of their own scheme, and a powerful array of
the Cheshire gentry and Weaver Navigation Trustees
appearing on behalf of the others,—and the whipping-up
of votes, in which the Duke of Bridgewater and Earl
Gower worked their influence with the Whig party to
good purpose.

Brindley's plan was, on the whole, considered the best.
It was the longest and the most circuitous, but it appeared
calculated to afford the largest amount of accommodation
to the public.   It would pass through important districts,
urgently in need of an improved communication with
the port of Liverpool on the one hand, and with Hull on
the other.   But it was not so much the connection of
those ports with each other that was needed, as a more
convenient means of communication between them and the
Staffordshire manufacturing districts ; and the Grand
Trunk system—somewhat in the form of a horse-shoe, with
the Potteries lying along its extreme convex part — promised effectually to answer this purpose, and to open up
a ready means of access to the coast on both sides of the
island.

A glance at the course of the proposed line will show
its great importance.   Starting from the Duke's canal at
Preston-on-the-Hill, near Runcorn, it passed southwards
by Northwich and Middlewich, through the great salt-

manufacturing districts of Cheshire, to the summit at Harecastle. It was alleged that the difficulties presented by the long tunnel at that point were so great that it could never be the intention of the projectors of the canal to carry their "chimerical idea," as it was called, into effect. Brindley however insisted, not only that the tunnel was practicable, but that, if the necessary powers were granted, he would certainly execute it.* Descending from the summit level into the valley of the Trent, the canal proceeded southwards through the Pottery districts, passing close to Burslem, Hanley, Stoke, and Lane End. It then passed onward, still south, by Trentham, Stone, and Shutborough, to Haywood, where it joined the canal projected to unite the Severn with the Mersey. Still following the valley of the Trent, the canal near Rugeley, turning sharp round, proceeded in a north-easterly direction, nearly parallel with the river, passing Burton and Ashton, to a junction with the main stream at Wilden Ferry, a little above where the Derwent falls into the Trent near Derby. From thence there was a clear line of navigation, by Nottingham, Newark, and Gainsborough, to the Humber. Provided this admirable project could be carried out, it appeared likely to meet all the necessities of the case. Ample evidence was given in support of the allegations of its promoters; and the result was, that Parliament threw out the bills promoted by the

---

* In one of the many angry pamphlets published at the time, the 'Supplement to a pamphlet entitled Seasonable Considerations on a Navigable Canal intended to be cut from the Trent to the Mersey,' &c., the following passage occurs : " When our all is at stake, these gentlemen [the promoters of the Grand Trunk Canal] must not be surprised at bold truths. We conceive more favourably of their *understanding* than of their *motive;* we cannot suspect them of entertaining the chimerical idea of cutting through *Hare Castle!* We rather believe that they are desirous of cutting their canal at both ends, and of leaving the middle for the project of a future day. Are these projectors *jealous* of their *honour?* Let them adopt a clause (which reason and justice strongly enforce) to restrain them from meddling with *either end* till they have finished the *great trunk.* This, and this alone, will shield them from suspicion.

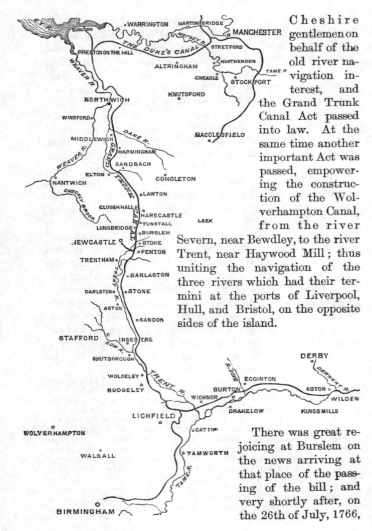

Cheshire gentlemen on behalf of the old river navigation interest, and the Grand Trunk Canal Act passed into law. At the same time another important Act was passed, empowering the construction of the Wolverhampton Canal, from the river Severn, near Bewdley, to the river Trent, near Haywood Mill; thus uniting the navigation of the three rivers which had their termini at the ports of Liverpool, Hull, and Bristol, on the opposite sides of the island.

There was great rejoicing at Burslem on the news arriving at that place of the passing of the bill; and very shortly after, on the 26th of July, 1766,

the works were formally commenced by Josiah Wedgwood
on the declivity of Bramhills, in a piece of land within a
few yards of the bridge which crosses the canal at that
place. Brindley was present at the ceremony, when due
honours were paid him by the assembled potters. After
Mr. Wedgwood had cut the first sod, many of the leading
persons of the neighbourhood followed his example, putting
their hand to the work by turns, and each cutting a turf
or wheeling a barrow of earth in honour of the occasion.
It was, indeed, a great day for the Potteries, as the event
proved. In the afternoon a sheep was roasted whole in
Burslem market-place, for the good of the poorer class of
potters; a *feu de joie* was fired in front of Mr. Wedgwood's house, and sundry other demonstrations of local
rejoicing wound up the day's proceedings.

Wedgwood was of all others the most strongly impressed
with the advantages of the proposed canal. He knew and
felt how much his trade had been hindered by the defective
communications of the neighbourhood, and to what extent
it might be increased provided a ready means of transit to
Liverpool, Hull, and Bristol could be secured; and, con-
fident in the accuracy of his anticipations, he proceeded to
make the purchase of a considerable estate in Shelton,
intersected by the canal, on the banks of which he built
the celebrated Etruria — the finest manufactory of the
kind up to that time erected in England, alongside of
which he built a mansion for himself and cottages for his
workpeople. He removed his works thither from Burslem,
partially in 1769, and wholly in 1771, shortly before the
works of the canal had been completed.

The Grand Trunk was the most formidable undertaking
of the kind that had yet been attempted in England. Its
whole length, including the junctions with the Birmingham Canal and the river Severn, was 139½ miles. In conformity with Brindley's practice, he laid out as much of
the navigation as possible upon a level, concentrating the
locks in this case at the summit, near Harecastle, from

which point the waters fell in both directions, north and south. Brindley's liking for long flat reaches of dead water made him keep clear of rivers as much as possible. He likened water in a river flowing down a declivity, to a furious giant running along and overturning everything; whereas (said he) "if you lay the giant flat upon his back, he loses all his force, and becomes completely passive, whatever his size may be." Hence he contrived that from Middlewich, a distance of seventeen miles, to the Duke's Canal at Preston Brook, there should not be a lock; but goods might be conveyed from the centre of Cheshire to Manchester, for a distance of about seventy miles, along one uniform water level. He carried out the same practice, in like manner, on the Trent side of Harecastle, where he laid out the canal in as many long lengths of dead water as possible.

The whole rise of the canal from the level of the Mersey, including the Duke's locks at Runcorn, to the summit at Harecastle, is 395 feet; and the fall from thence to the Trent at Wilden is 288 feet 8 inches. The locks of the Grand Trunk proper, on the northern side of Harecastle, are thirty-five, and on the southern side forty. The dimensions of the canal, as originally constructed, were twenty-eight feet in breadth at the top, sixteen at the bottom, and four and a half feet in depth; but from Wilden to Burton, and from Middlewich to Preston-on-the-Hill, it was thirty-one feet broad at the top, eighteen at the bottom, and five and a half feet deep, so as to be navigable by large barges; and the locks at those parts of the canal were of correspondingly large dimensions. The width was afterwards made uniform throughout. The canal was carried over the river Dove on an aqueduct of twenty-three arches, approached by an embankment on either side—in all a mile and two furlongs in length. There were also aqueducts over the Trent, which it crosses at four different points—one of these being of six arches of twenty-one feet span each—and over the Dane and other smaller streams.

The number of minor aqueducts was about 160, and of road-bridges 109.

But the most formidable works on the canal were the tunnels, of which there were five—the Harecastle, 2880 yards long; the Hermitage, 130 yards; the Barnton, 560 yards; the Saltenford, 350 yards; and the Preston-on-the-Hill, 1241 yards. The Harecastle tunnel (subsequently duplicated by Telford) was constructed only nine feet wide and twelve feet high;* but the others were seventeen feet four inches high, and thirteen feet six inches wide. The most extensive ridge of country to be penetrated was at Harecastle, involving by far the most difficult work in the whole undertaking. This ridge is but a continuation of the high ground, forming what is called the "back-bone of England," which extends in a south-westerly direction from the Yorkshire mountains to the Wrekin in Shropshire. The flat county of Cheshire, which looks almost as level as a bowling-green when viewed from the high ground near New Chapel, seems to form a deep bay in the land, its innermost point being almost immediately under the village of Harecastle; and from thence to the valley of the Trent the ridge is at the narrowest. That Brindley was correct in determining to form his tunnel at this point has since been confirmed by the survey of Telford, who there constructed his parallel tunnel for the same canal, and still more recently by the engineers of the North Staffordshire Railway, who have also formed their railway tunnel almost parallel with the line of both canals.

When Brindley proposed to cut a navigable way under

---

* Brindley's tunnel had only space for a narrow canal-boat to pass through, and it was propelled by the tedious and laborious process of what is called "legging." It still continues to be worked in the same way, while horses haul the boats through the whole length of Telford's wider tunnel. The men who "leg" the boat, literally kick it along from one end to the other. They lie on their backs on the boat-cloths, with their shoulders resting against some package, and propel it along by means of their feet pressing against the top or sides of the tunnel.

this ridge, it was declared to be chimerical in the extreme. The defeated promoters of the rival projects continued to make war upon it in pamphlets, and in the exasperating language of mock sympathy proclaimed Brindley's proposed

Northern Entrance of Harecastle Tunnels.*

tunnel to be " a sad misfortune," † inasmuch as it would utterly waste the capital raised by the subscribers, and end in the inevitable ruin of the concern. Some of the small local wits spoke of it as another of Brindley's " Air Castles;" but the allusion was not a happy one, as his first

---

* The smaller opening into the hill on the right-hand of the view is Brindley's tunnel; that on the left is Telford's, executed some forty years since. Harecastle church and village occupy the ground over the tunnel entrances.

† 'Seasonable Considerations,' &c.; Canal pamphlet dated 1766.

" castle in the air," despite all prophecies to the contrary, had been built, and continued to stand firm at Barton : and judging by the issue of that undertaking, it was reasonable to infer that he might equally succeed in this, difficult though it was on all hands admitted to be.

The Act was no sooner passed than Brindley set to work to execute the impossible tunnel. Shafts were sunk from the hill-top at different points down to the level of the intended canal. The stuff was drawn out of the shafts in the usual way by horse-gins ; and so long as the water was met with in but small quantities, the power of windmills and watermills working pumps over each shaft was sufficient to keep the excavators at work. But as the miners descended and cut through the various strata of the hill on their downward progress, water was met with in vast quantities ; and here Brindley's skill in pumping machinery proved of great value. The miners were often drowned out, and as often set to work again by his mechanical skill in raising water. He had a fire-engine, or atmospheric steam-engine, of the best construction possible at that time, erected on the top of the hill, by the action of which great volumes of water were pumped out night and day.

This abundance of water, though it was a serious hinderance to the execution of the work, was a circumstance on which Brindley had calculated, and indeed depended, for the supply of water for the summit level of his canal. When the shafts had been sunk to the proper line of the intended waterway, the excavation then proceeded in opposite directions, to meet the other driftways which were in progress. The work was also carried forward at both ends of the tunnel, and the whole line of excavation was at length united by a continuous driftway—it is true, after long and expensive labour—when the water ran freely out at both ends, and the pumping apparatus on the hilltop was no longer needed. At a general meeting of the Company, held on the 1st October, 1768, after the works had been in progress about two years, it appeared from the report of the Committee that four hundred and nine yards of the tunnel were cut and vaulted, besides the vast excavations at either end for the purpose of reservoirs ; and the Committee expressed their opinion that the work would be finished without difficulty.

Active operations had also been in progress at other parts of the canal. About six hundred men in all were employed, and Brindley went from point to point superintending and directing their labours. A Burslem correspondent, in September, 1767, wrote to a distant friend thus :—" Gentlemen come to view our eighth wonder of the world, the subterraneous navigation, which is cutting by the great Mr. Brindley, who handles rocks as easily as you would plum-pies, and makes the four elements subservient to his will. He is as plain a looking man as one of the boors of the Peak, or as one of his own carters ; but when he speaks, all ears listen, and every mind is filled with wonder at the things he pronounces to be practicable. He has cut a mile through bogs, which he binds up, embanking them with stones which he gets out of other parts of the navigation, besides about a quarter of a mile into the hill Yelden, on the side of which he has a pump worked by water, and a stove, the fire of which sucks through a pipe the damps that would annoy the men who are cutting towards the centre of the hill. The clay he cuts out serves for bricks to arch the subterraneous part, which we heartily wish to see finished to Wilden Ferry, when we shall be able to send Coals and Pots to London, and to different parts of the globe."

In the course of the first two years' operations, twenty-two miles of the navigation had been cut and finished, and it was expected that before eighteen months more had elapsed the canal would be ready for traffic by water between the Potteries and Hull on the one hand, and Bristol on the other. It was also expected that by the same time the canal would be ready for traffic from the north end of Harecastle Tunnel to the river Mersey. The execution of the tunnel, however, proved so tedious and difficult, and the excavation and building went on so slowly, that the Committee could not promise that it would be finished in less than five years from that time. As it was, the completion of the Harecastle Tunnel occupied nine years more; and it was not finished until the year 1777, by which time the great engineer had finally rested from all his labours.

It is scarcely necessary to describe the benefits which the canal conferred upon the inhabitants of the districts through which it passed. As we have already seen, Stafford-shire and the adjoining counties had been inaccessible during the chief part of each year. The great natural wealth which they contained was of little value, because it could with difficulty be got at; and even when reached, there was still greater difficulty in distributing it. Coal could not be worked at a profit, the price of land-carriage so much restricting its use, that it was placed altogether beyond the reach of the great body of consumers.

It is difficult now to realise the condition of poor people situated in remote districts of England less than a century ago. In winter time they shivered over scanty wood-fires, for timber was almost as scarce and as dear as coal. Fuel was burnt only at cooking times, or to cast a glow about the hearth in the winter evenings. The fireplaces were little apartments of themselves, sufficiently capacious to enable the whole family to sit within the wide chimney, where they listened to stories or related to each other the events of the day. Fortunate were the villagers who lived hard by a bog or a moor, from which they could cut peat or turf at will. They ran all risks of ague and fever in summer, for the sake of the ready fuel in winter. But in places remote from bogs, and scantily timbered, existence was scarcely possible; and hence the settlement and culti-vation of the country were in no slight degree retarded until comparatively recent times, when better communica-tions were opened up.

I.   T

When the canals were made, and enabled coals to be readily conveyed along them at comparatively moderate rates, the results were immediately felt in the increased comfort of the people. Employment became more abundant, and industry sprang up in their neighbourhood in all directions. The Duke's canal, as we have seen, gave the first great impetus to the industry of Manchester and that district. The Grand Trunk had precisely the same effect throughout the Pottery and other districts of Staffordshire; and their joint action was not only to employ, but actually to civilize the people. The salt of Cheshire could now be manufactured in immense quantities, readily conveyed away, and sold at a comparatively moderate price in all the midland districts of England. The potters of Burslem and Stoke, by the same mode of conveyance, received their gypsum from Northwich, their clay and flints from the seaports now directly connected with the canal, and returned their manufactures by the same route. The carriage of all articles being reduced to about one-fourth of their previous rates,* articles of necessity and comfort, such as had formerly been unknown except amongst the wealthier

---

* The following comparison of the rates per ton at which goods were conveyed by land-carriage before the opening of the Grand Trunk Canal, and those at which they were subsequently carried by it, will show how great was the advantage conferred on the country by the introduction of navigable canals:—"The cost of carrying a ton of goods from Liverpool to Etruria, the centre of the Staffordshire Potteries, by land-carriage, was 50s.; the Trent and Mersey reduced it to 13s. 4d. The land-carriage from Liverpool to Wolverhampton was 5l. a ton; the canal reduced it to 1l. 5s. The land-carriage from Liverpool to Birmingham, and also to Stourport, was 5l.

a ton; the canal reduced both to 1l. 10s. . . . . . . Thus the cost of inland transport was reduced, on the average, to about one-fourth of the rate paid previous to the introduction of canal navigation. The advantages were enormous: wheat, for example, which formerly could not be conveyed a hundred miles, from corn-growing districts to the large towns and manufacturing districts, for less than 20s. a quarter, could be conveyed for about 5s. a quarter. These facts show how great was the service conferred on the country by Brindley and the Duke of Bridgewater." — Baines's 'History of the Commerce and Town of Liverpool.'

classes, came into common use amongst the people. Employment increased, and the difficulties of subsistence diminished. Led by the enterprise of Wedgwood and others like
him, new branches of industry sprang up, and the manufacture of earthenware, instead of being insignificant and
comparatively unprofitable, which it was before his time,
became a staple branch of English trade. Only about
ten years after the Grand Trunk Canal had been opened,
Wedgwood stated in evidence before the House of Commons, that from 15,000 to 20,000 persons were then employed in the earthenware-manufacture alone, besides the
large number of labourers employed in digging coals for
their use, and the still larger number occupied in providing materials at distant parts, and in the carrying and
distributing trade by land and sea. The annual import
of clay and flints into Staffordshire at that time was from
fifty to sixty thousand tons ; and yet, as Wedgwood truly
predicted, the trade was but in its infancy. The tonnage
outwards and inwards at the Potteries is now upwards of
three hundred thousand tons a-year.

The moral and social influences exercised by the canals
upon the Pottery districts were not less remarkable.
From a half-savage, thinly-peopled district of some 7000
persons in 1760, partially employed and ill-remunerated,
we find them increased, in the course of some twenty-five
years, to about treble the population, abundantly employed, prosperous, and comfortable.* Civilization is
doubtless a plant of very slow growth, and does not necessarily accompany the rapid increase of wealth. On the
contrary, higher earnings, without improved morale, may
only lead to wild waste and gross indulgence. But the
testimony of Wesley to the improved character of the
population of the Pottery district in 1781, within a few
years after the opening of Brindley's Grand Trunk Canal,

---

* The population of the same district in 1861 was found to be upwards
of 120,000.

is so remarkable, that we cannot do better than quote it
here ; and the more so, as we have already given the account of his first visit in 1760, on the occasion of his
being pelted. " I returned to Burslem," says Wesley ;
" how is the whole face of the country changed in about
twenty years ! Since which, inhabitants have continually
flowed in from every side. Hence the wilderness is literally
become a fruitful field. Houses, villages, towns, have
sprung up, and the country is not more improved than
the people."

CHAPTER VII.

BRINDLEY'S LAST CANALS — HIS DEATH AND CHARACTER.

IT is related of Brindley that, on one occasion, when giving evidence before a Committee of the House of Commons, in which he urged the superiority of canals over rivers for purposes of inland navigation, the question was asked by a member, " Pray, Mr. Brindley, what then do you think is the use of navigable rivers?" " To make canal navigations, to be sure," was his instant reply. It is easy to understand the gist of the engineer's meaning. For purposes of trade he regarded regularity and certainty of communication as essential conditions of any inland navigation; and he held that neither of these could be relied upon in the case of rivers, which are in winter liable to interruption by floods, and in summer by droughts. In his opinion, a canal, with enough of water always kept banked up, or locked up where the country would not admit of the level being maintained throughout, was absolutely necessary to satisfy the requirements of commerce. Hence he held that one of the great uses of rivers was to furnish a supply of water for canals. It was only another illustration of the " nothing like leather " principle; Brindley's head being so full of canals, and his labours so much confined to the making of canals, that he could think of little else.

In connection with the Grand Trunk—which proved, as Brindley had anticipated, to be the great aorta of the canal system of the midland districts of England—numerous lines were projected and afterwards carried out under our engineer's superintendence. One of the most important of these was the Wolverhampton Canal, connecting the Trent with the Severn, and authorised in the same

year as the Grand Trunk itself. It is now known as
the Staffordshire and Worcestershire Canal, passing close
to the towns of Wolverhampton and Kidderminster, and
falling into the Severn at Stourport. This branch opened
up several valuable coal-fields, and placed Wolverhampton
and the intermediate districts, now teeming with population and full of iron manufactories, in direct connection
with the ports of Liverpool, Hull, and Bristol. Two years
later, in 1768, three more canals, laid out by Brindley,
were authorised to be constructed: the Coventry Canal
to Oxford, connecting the Grand Trunk system by Lichfield with London and the navigation of the Thames; the
Birmingham Canal, which brought the advantages of inland navigation to the very doors of the central manufacturing town in England; and the Droitwich Canal, to
connect that town by a short branch with the river Severn.
In the following year a further Act was obtained for a
canal laid out by Brindley, from Oxford to the Coventry
Canal at Longford, eighty-two miles in length.

These were highly important works; and though they
were not all carried out strictly after Brindley's plans,
they nevertheless formed the groundwork of future Acts,
and laid the foundations of the midland canal system.
Thus, the Coventry Canal was never fully carried out
after Brindley's designs; a difference having arisen between the engineer and the Company during the progress
of the undertaking, in consequence, as is supposed, of the
capital provided being altogether inadequate to execute
the works considered by Brindley as indispensable. He
probably foresaw that there would be nothing but difficulty, and very likely there might be discredit attached
to himself by continuing connected with an undertaking the
proprietors of which would not provide him with sufficient
means for carrying it forward to completion; and though
he finished the first fourteen miles between Coventry
and Atherstone, he shortly after gave up his connection
with the undertaking, and it remained in an unfinished

state for many years, in consequence of the financial difficulties in which the Company had become involved through the insufficiency of their capital. The connection of the Coventry Canal with the Grand Trunk was afterwards completed, in 1785, by the Birmingham and Fazeley and Grand Trunk Companies conjointly, and the property eventually proved of great value to all parties concerned.

The Droitwich Canal, though only a short branch five and a half miles in length, was a very important work, opening up as it did an immense trade in coal and salt between Droitwich and the Severn. The works of this navigation were wholly executed by Brindley, and are considered superior to those of any others on which he was engaged. Whilst residing at Droitwich, we find our engineer actively engaged in pushing on the subscription to the Birmingham Canal, the capital of which was taken slowly. Matthew Boulton, of Birmingham, was one of the most active promoters of the scheme, and Josiah Wedgwood also bestirred himself in its behalf. In a letter written by him about this time, we find him requesting one of his agents to send out plans to gentlemen whom he names, in the hope of completing the subscription-list.* Brindley did not live to finish the Bir-

---

* The letter is so characteristic of Josiah Wedgwood that we here insert it at length, as copied from the original in the possession of Mr. Mayer of Liverpool :—

" Burslem, 12th July, 1769

" Dear Sir,—I should have wrote to you about young Wilson, but the multiplicity of branches you wrote me he was expected to learn, made me despair of teaching him any. Pray give my compliments to his father, and if he chooses to have his son to learn to be a warehouseman and book-keeper, which is quite sufficient and better than more for any one person, I will learn him those in the best manner; but, even then, Mr. Wilson must not expect him to be set on the top of a ladder without set-ting his feet upon the lowermost steps; and unless he will let the Boy pursue that method, I would not be concerned with him on any account. I will not attempt to teach him any more trades; it would injure the Boy, and do me no good. If he has a mind at his leisure time to amuse himself with drawing I have no objection, and would encourage him in it, as an innocent amusement, and what may be of use to him, but would not make this a branch of his business. If the business I propose is too humble for Mr. Wilson's son, I would not by any means have him accept of it.

" Mr. Brindley desires you'll send 30 plans to each of the undermentioned Gentlemen by the first Waggons, and let us know when they are sent, as we shall advertise them in several of the Country papers: Mr. Walker, of Oxford, Steward to D. of Marlbro'—you may perhaps get

mingham Canal; it was carried out by his successors,—partly by his pupil, Mr. Whitworth, and partly by Smeaton and Telford. Brindley's plan was, as usual, to cut the canal as flat as possible, to avoid the necessity for lockage; but his successors, in order to relieve the capital expenditure, as they supposed, constructed it with a number of locks to carry it over the summit at Smithwick. Shortly after its opening, however, the Company found reason to regret their rejection of Mr. Brindley's advice, and they lowered the summit by cutting a tunnel, as he had originally recommended, thereby incurring an extra expense of about 30,000*l.*

Another of Brindley's canals, authorised in 1769, was that between Chesterfield and the river Trent, at Stockwith, about forty-six miles in length, intended for the transport of coal, lime, and lead from the rich mineral districts of Derbyshire, and the return trade of deals, corn, and groceries to the same districts. It would appear that Mr. Grundy, another engineer, of considerable reputation in his day, was consulted about the project, and that he advised a much more direct route than that pointed out by Brindley, who looked to the accommodation of the existing towns, rather than shortness of route, as the main thing to be provided for. Brindley, in this respect, took very much the same view in laying out his canals as was afterwards taken by George Stephenson—a man whom he resembled in many respects—in laying out railways. He would rather go *round* an obstacle in the shape of an elevated range of country, than go *through* it, especially if in going round and avoiding expense he could accommodate a number of towns and villages. Besides, by avoiding the hills and following the course of the valleys,

---

them sent from Marlbro' house; Mr. Dudley, Attorney - at - Law, Coventry; Mr. Richardson, Silversmith, in Chester; Mr. Perry, Wolverhampton.

"We shall send you this week end, double salts, creams, pott'y potts, table plates of all sorts, sallad dishes, covered do., pierced desert plates, &c. We cannot get Sadler to send us any pierced desert ware, &c. Pay Addison 5 or 6 guineas.

"Yours, &c., J W.'

along which the population usually lies, he avoided expense
of construction and secured flatness of canal; just as
Stephenson secured flatness of railway gradient. Although
the length of canal to be worked was longer, yet the
cost of tunnelling and lockage was avoided. The population of the district was also fully accommodated, which
could not have been accomplished by the more direct
route through unpopulated districts or under barren hills.
The proprietors of the Chesterfield Canal concurred in
Brindley's view, adopting his plan in preference to
Grundy's, and it was accordingly carried into effect. This
navigation was, nevertheless, a work of considerable difficulty, proceeding, as it did, across a very hilly country,
the summit tunnel at Hartshill being 2850 yards in extent. Like many of Brindley's other works projected
about this time, it was finished by his brother-in-law, Mr.
Henshall, and opened for traffic several years after the
great engineer's death.*

The whole of these canals were laid out by Brindley,
though they were not all executed by him, nor precisely
after his plans. No record of any kind has been preserved of the manner in which the works were carried
out. Brindley himself made few reports, and these merely
stated results, not methods; yet he had doubtless many
formidable difficulties to encounter, and must have overcome them by the adoption of those ingenious expedients,
varying according to the circumstances of each case, in
which he was always found so fertile. He had no trea-

---

* The following were the canals laid out and principally executed by
Brindley:—

		Miles.	furl.	chains.
The Duke's Canals { Worsley to Manchester		10	2	0
Longford Bridge to the Mersey below Runcorn		24	1	7
Grand Trunk Canal Proper, from Wilden Ferry to Preston Brook		88	7	9
Wolverhampton Canal		46	4	0
Coventry ,,		36	7	8
Birmingham ,,		24	2	0
Droitwich ,,		5	4	9
Oxford ,,		82	7	3
Chesterfield ,,		46	0	0

sury of past experience, as recorded in books, to consult, for he could scarcely read English; and certainly he could neither read French nor Italian, in which languages the only engineering works of any value were then written; nor had he any store of native experience to draw from, he himself being the first English canal engineer of eminence, and having all his methods and expedients to devise for himself.

It would doubtless have been most interesting could we have had some authentic record of this strong original man's struggles with opposition and difficulty, and the means by which he contrived not only to win persons of high station to support him with their influence but also with their purses, at a time when money was comparatively a much rarer commodity than it is now. "That want of records, journals, and memoranda," says Mr. Hughes, "which is ever to be deplored when we seek to review the progress of engineering works, is particularly felt when we have to look back upon those undertakings which first called for the exercise of engineering skill in many new and untried departments. In Brindley's day, the entire absence of experience derived from former works, the obscure position which the engineer occupied in the scale of society, the imperfect communication between the profession in this country and the engineers and works of other countries, and, lastly, the backward condition of all the mechanical arts and of the physical sciences connected with engineering, may all be ranked in striking contrast with the vast appliances which are placed at the command of modern engineers." *

Moreover, the great canal works upon which Brindley was engaged during the later part of his career, were as yet scarcely appreciated as respects the important influences which they were calculated to exercise upon society

---

* 'Memoir,' by Samuel Hughes, C.E.   Weale's 'Quarterly Papers, 1844.

at large. The only persons who seem to have regarded them with interest were far-sighted men like Josiah Wedgwood, who saw in them the means not only of promoting the trade of his own county, but of developing the rich natural resources of the kingdom, and diffusing amongst the people the elements of comfort, intelligence, and civilization. The literary and scientific classes as yet took little or no interest in them. The most industrious and observant literary man of the age, Dr. Johnson, though he had a word to say upon nearly every subject, never so much as alluded to them, though all Brindley's canals were finished in Johnson's lifetime, and he must have observed the works in progress when passing on his various journeys through the midland districts. The only reference which he makes to the projects set on foot for opening up the country by means of better roads, was to the effect, that whereas there were before cheap places and dear places, now all refuges were destroyed for elegant or genteel poverty.

Before leaving this part of the subject, it is proper to state that during the latter part of Brindley's life, whilst canals were being projected in various directions, he was, on many occasions, called upon to give his opinion as to the plans which other engineers had prepared. Among the most important of the new projects on which he was thus consulted were, the Leeds and Liverpool Canal; the improvement of the navigation of the Thames to Reading; the Calder Navigation; the Forth and Clyde Canal; the Salisbury and Southampton Canal; the Lancaster Canal; and the Andover Canal. Many of these schemes were of great importance in a national point of view. The Leeds and Liverpool Canal, for instance, brought the whole manufacturing district of Yorkshire along the valley of the Aire into communication with Liverpool and the intermediate districts of Lancashire. The advantages of this navigation to Leeds, Bradford, Keighley, and the neighbouring towns, are felt to this day, and their extraordinary prosperity is doubtless in no small degree attributable to the facilities which the canal has provided for the ready conveyance of raw materials and manufactured produce between those places and the towns and sea-ports of the west. Brindley surveyed and laid out the whole line of this navigation, 130 miles in length, and he framed the estimate on which the Company proceeded to Parliament for their bill. On the passing of the Act in 1768-9, the Directors appointed him their engineer; but, being almost overwhelmed with other business at the time, and feeling that he could not give the proper degree of personal attention to carrying out so extensive an undertaking, he was under the necessity of declining the appointment. The works were immediately begun at both ends of the canal, and portions were speedily made use of; but the difficulty and expensiveness of the remaining works greatly delayed their execution, and the canal was not finished until the year 1816. Twenty miles, extending from near Bingley to the neighbourhood of Bradford, were opened on 21st March, 1774. A correspondent of 'Williamson's Liverpool Advertiser' thus describes the opening : " From Bingley and about three miles down, the noblest works of the kind that perhaps are to be found in the universe are exhibited, namely, a five-fold, a three-fold, a two-fold, and a single lock, making together a fall of 120 feet; a large aqueduct-bridge of seven arches over the river Aire, and an aqueduct on a large embankment over Shipley valley. Five boats of burden passed the grand lock, the first of which descended through a fall of sixty-six feet in less than twenty-nine minutes. This much wished-for-event was welcomed with ringing of bells, a band of music, the firing of guns by the neighbouring militia, the shouts of spectators, and all the marks of satisfaction that so important an acquisition merits." On the 21st October of the same year the following paragraph appeared :— " The Liverpool end of the canal was opened from Liverpool to Wigan on Wednesday, the 19th instant, with great festivity and rejoicings. The water had been led into the basin the evening before.   At nine A.M. the proprietors sailed up the canal in their barge, preceded by another with music, colours flying, &c., and returned to Liverpool about one.   They were saluted with two royal salutes of twenty-one guns each, besides the swivels on board the boats, and welcomed with the repeated shouts of the numerous crowds assembled on the banks, who made a most cheerful and agreeable sight. The gentlemen then adjourned to a tent, on the quay, where a cold collation was set out for themselves and their friends.   From thence they went in procession to George's coffee-house, where an elegant dinner was provided.   The workmen, 215 in number, walked first, with their tools on their shoulders, and cockades in their hats, and were afterwards plentifully regaled at a dinner provided for them.   The bells rang all day, and the greatest joy and order prevailed on the occasion."

Brindley being now the recognised head of his profession, and the great authority on all questions of navigation, he was, in 1770, employed by the Corporation of London to make a survey of the Thames above Battersea, with the object of having it improved for purposes of navigation.   As usual, Brindley strongly recommended the construction of a canal in preference to carrying on the navigation by the river, where it was liable to be interrupted by the tides and floods, or by the varying deposits of silt in the shallow places.   In his first report to the Common Council, dated the 16th of June, 1770, he pointed out that the cost of hauling the barges was greatly in favour of the canal.   For example, he stated that the expense of taking a vessel of 100 or 120 tons from Isleworth to Sunning, and back again to Isleworth, was 80*l*., and sometimes more ; whilst the cost by the canal would only be 16*l*.   The saving in time would be still greater, for the double voyage might easily be performed in fifteen hours ; whereas by the river the boats were sometimes three weeks in going up, and almost as much in coming down. He estimated that there would be a saving to the public of at least 64*l.* on every voyage, besides the saving of time in performing it. After making a further detailed examination of the district, and maturing his views on the whole subject, he sent in a report, accompanied by a profile of the river about seven feet long, which is still to be seen amongst the records of the Corporation of London. His plan was not, however, carried out; the proposal to construct a canal parallel with the Thames having been abandoned so soon as the Grand Junction Canal was undertaken.

These and numerous other schemes in various parts of the country—at Stockton, at Leeds, at Cambridge, at Chester, at Salisbury and Southampton, at Lancaster, and in Scotland—fully occupied the attention of Brindley; in addition to which, there was the personal superintendence which he must necessarily give to the canals in active progress, and for the execution of which he was responsible. In fact, there was scarcely a design of a canal navigation set on foot throughout the kingdom during the later years of his life, on which he was not consulted, and the plans of which he did not entirely make, revise, or improve.

In addition to his canal works, Brindley was also consulted as to the best means of draining the low lands in different parts of Lincolnshire, and the Great Level in the neighbourhood of the Isle of Ely. He supplied the corporation of Liverpool with a plan for cleansing the docks and keeping them clear of mud, which is said to have proved very effective; and he pointed out to them an economical method of building walls against the sea without mortar, which long continued to be employed with complete success. In such cases he laid his plans freely open to the public, not seeking to secure them by patent, nor shrouding his proceedings in any mystery. He was perfectly open with professional men, harbouring no petty jealousy of rivals.   His pupils had free access to all his methods, and he took a pride in so training them that they should reflect credit on the engineer's profession, then rising into importance, and be enabled, after he left the scene, to carry on those great industrial enterprises which he probably foresaw clearly enough in England's future.

It will be observed, from what we have said, that Brindley's engagements as an engineer extended over a very wide district.   Even before his employment by the Duke of Bridgewater, he was under the necessity of travelling great distances to fit up water-mills, pumping-engines, and manufacturing machinery of various kinds, in the counties of Stafford, Cheshire, and Lancashire.   But when he had been appointed to superintend the construction of the Duke's canals, his engagements necessarily became of a still more engrossing character, and he had very little leisure left to devote to the affairs of private life.   He lived principally at inns, in the immediate neighbourhood of his work ; and though his home was at Leek, he sometimes did not visit it for weeks together.

Brindley had very little time for friendship, and still less for courtship.   Nevertheless, he did contrive to find time for marrying, though at a comparatively advanced period of his life.   In laying out the Grand Trunk Canal, he was necessarily brought into close connection with Mr. John Henshall, of the Bent, near New Chapel, land-surveyor, who assisted him in making the survey.   He visited Henshall at his house in September, 1762, and then settled with him the preliminary operations.   During his visits Brindley seems to have taken a special liking for Mr. Henshall's daughter Anne, then a girl at school, and when he went to see her father, he was accustomed to take a store of gingerbread for Anne in his pocket.   She must have been a comely girl, judging by the portrait of her as a woman, which we have seen.

In due course of time, the liking ripened into an attach-
ment; and shortly after the girl had left school, at the age
of only nineteen, Brindley proposed to her, and was
accepted.   By that time he was close upon his fiftieth
year, so that the union may possibly have been quite as
much a matter of convenience as of love on his part.   He
had now left the Duke's service for the purpose of entering
upon the construction of the Grand Trunk Canal, and with
that object resolved to transfer his home to the immediate
neighbourhood of Harecastle, as well as of his colliery at
Golden Hill.   Shortly after the marriage, the old mansion
of Turnhurst fell vacant, and Brindley with his young wife
became its occupants.   The marriage took place on the 8th
December, 1765, in the parish church of Wolstanton,
Brindley being described in the register as "of the parish
of Leek, engineer;" but from that time until the date of
his death his home was at Turnhurst.

Brindley's House at Turnhurst.

The house at Turnhurst was a comfortable, roomy, old-fashioned dwelling, with a garden and pleasure-ground behind, and a little lake in front. It was formerly the residence of the Bellot family, and is said to have been the last mansion in England in which a family fool was maintained. Sir Thomas Bellot, the last of the name, was a keen sportsman, and the panels of several of the upper rooms contain pictorial records of some of his exploits in the field. In this way Sir Thomas seems to have squandered his estate, and it shortly after became the property of the Alsager family, from whom Brindley rented it. A little summer-house, standing at the corner of the outer court-yard, is still pointed out as Brindley's office, where he sketched his plans and prepared his calculations. As for his correspondence, it was nearly all conducted, subsequent to his marriage, by his wife, who, notwithstanding her youth, proved a most clever, useful, and affectionate partner.

Turnhurst was conveniently near to the works then in progress at Harecastle Tunnel, which was within easy walking distance, whilst the colliery at Golden Hill was only a few fields off. From the elevated ground at Golden Hill, the whole range of high ground may be seen under which the tunnel runs—the populous Pottery towns of Tunstall and Burslem filling the valley of the Trent towards the south. At Golden Hill, Brindley carried out an idea which he had doubtless brought with him from Worsley. He and his partners had an underground canal made from the main line of the Harecastle Tunnel into their coal-mine, about a mile and a half in length; and by that tunnel the whole of the coal above that level was afterwards worked out, and conveyed away for sale in the Pottery and other districts, to the great profit of the owners and much to the convenience of the public.

These various avocations involved a great amount of labour as well as anxiety, and probably considerable tear and wear of the vital powers. But we doubt whether mere hard work ever killed any man, or whether Brindley's labours, extraordinary though they were, would have shortened his life, but for the far more trying condition of the engineer's vocation—irregular living, exposure in all weathers, long fasting, and then, perhaps, heavy feeding when the nervous system was exhausted, together with habitual disregard of the ordinary conditions of physical health. These are the main causes of the shortness of life of most of our eminent engineers, rather than the amount and duration of their labours. Thus the constitution becomes strained, and is ever ready to break down at the weakest place. Some violation of the natural laws more flagrant than usual, or a sudden exposure to cold or wet, merely presents the opportunity for an attack of disease which the ill-used physical system is found unable to resist.

Such an accidental exposure unhappily proved fatal to Brindley. While engaged one day in surveying a branch canal between Leek and Froghall, he got drenched near Ipstones, and went about for some time in his wet clothes. This he had often before done with impunity, and he might have done so again; but, unfortunately, he was put into a damp bed in the inn at Ipstones, and this proved too much for his constitution, robust though he naturally was. He became seriously ill, and was disabled from all further work. Diabetes shortly developed itself, and, after an illness of some duration, he expired at his house at Turnhurst, on the 27th of September, 1772, in the fifty-sixth year of his age, and was interred in the burying-ground at New Chapel, a few fields distant from his dwelling.

James Brindley was probably one of the most remarkable instances of self-taught genius to be found in the whole range of biography. The impulse which he gave to social activity, and the ameliorative influence which he exercised upon the condition of his countrymen, seem out of all proportion to the meagre intellectual culture which he had received in the course of his laborious and active career. We must not, however, judge him merely by the literary test.   It is true, he could scarcely read, and he was thus cut off, to his own great loss, from familiar intercourse with a large class of cultivated minds, living and dead ; for he could with difficulty take part in the conversation of educated men, and he was unable to profit by the rich stores of experience treasured up in books. Neither could he write, except with difficulty and inaccurately, as we have shown by the extracts above quoted from his note-books, which are still extant.

Brindley was, nevertheless, a highly-instructed man in many respects.   He was full of the results of careful observation, ready at devising the best methods of over-coming material difficulties, and possessed of a powerful and correct judgment in matters of business.   When any emergency arose, his quick invention and ingenuity, culti-vated by experience, enabled him almost at once unerringly to suggest the best means of providing for it.   His ability in this way was so remarkable, that those about him attri-buted the process by which he arrived at his conclusions rather to instinct than reflection — the true instinct of genius.   " Mr. Brindley," said one of his contemporaries, " is one of those great geniuses whom Nature sometimes rears by her own force, and brings to maturity without the necessity of cultivation.   His whole plan is admirable, and so well concerted that he is never at a loss ; for, if any difficulty arises, he removes it with a facility which appears so much like inspiration, that you would think Minerva was at his fingers' ends."

His mechanical genius was indeed most highly culti-vated.   From the time when he bound himself apprentice to the trade of a millwright—impelled to do so by the strong bias of his nature—he had been undergoing a course of daily and hourly instruction.   There was nothing to distract his attention, or turn him from pursuing his favourite study of practical mechanics.   The training of his inventive faculty and constructive skill was, indeed, a slow but a continuous process ; and when the time and the opportunity arrived for turning these to account—when the silk-throwing machinery of the Congleton mill, for instance, had to be perfected and brought to the point of effectively performing its intended work—Brindley was found able to take it in hand and carry out the plan, when even its own designer had given it up in despair. But it must also be remembered that this extraordinary ability of Brindley was in a great measure the result of close observation, pains-taking study of details, and the most indefatigable industry.

The same qualities were displayed in his improvements of the steam-engine, and his arrangements to economise power in the pumping of water from drowned mines. It was often said of his works, as was said of Columbus's discovery, "How easy! how simple!" but this was after the fact. Before he had brought his fund of experience and clearness of vision to bear upon a difficulty, every one was equally ready to exclaim "How difficult! how absolutely impracticable!" This was the case with his "castle in the air," the Barton Viaduct—such a work as had never before been attempted in England, though now any common mason would undertake it. It was Brindley's merit always to be ready with his simple, practical expedient; and he rarely failed to effect his purpose, difficult although at first sight its accomplishment might seem to be.

Like men of a similar stamp, Brindley had great confidence in himself and in his powers and resources. Without this, it had been impossible for him to have accomplished so much as he did. It is said that the King of France, hearing of his wonderful genius, and the works he had performed for the Duke of Bridgewater at Worsley, expressed a desire to see him, and sent a message inviting him to view the Grand Canal of Languedoc. But Brindley's reply was characteristic : " I will have no journeys to foreign countries," said he, " unless to be employed in surpassing all that has been already done in them."

His observation was remarkably quick. In surveying a

district, he rapidly noted the character of the country, the direction of the hills and the valleys, and, after a few journeys on horseback, he clearly settled in his mind the best line to be selected for a canal, which almost invariably proved to be the right one. In like manner he would estimate with great rapidity the fall of a brook or river while walking along the banks, and thus determined the height of his cuttings and embankments, which he afterwards settled by a more systematic survey. In these estimates he was rarely, if ever, found mistaken.

His brother-in-law, Mr. Henshall, has said of him, "when any extraordinary difficulty occurred to Mr. Brindley in the execution of his works, having little or no assistance from books or the labours of other men, his resources lay within himself. In order, therefore, to be quiet and uninterrupted whilst he was in search of the necessary expedients, he generally retired to his bed;* and he has been known to be there one, two, or three days, till he had attained the object in view. He would then get up and execute his design, without any drawing or model. Indeed, it was never his custom to make either, unless he was obliged to do it to satisfy his employers. His memory was so remarkable that he has often declared that he could remember, and execute, all the parts of the most complex machine, provided he had time, in his survey of it, to settle in his mind the several parts and their relations to each other. His method of calculating the powers of any machine invented by him was peculiar to himself. He worked the question for some time in his head, and then put down the results in figures. After this, taking it up again at that stage, he worked it further in his mind for a certain time, and set down the results as before. In the same way he

---

* The younger Pliny seems to have adopted almost a similar method · "Clausæ fenestræ manent. Mirè enim silentio et tenebris animus alitur. Ab iis quæ avocant abductus, et liber, et mihi relictus, non oculos animo sed animum oculis sequor, qui eadem quæ mens vident quoties non vident alia."—Epist. lib. ix., ep. 36.

still proceeded, making use of figures only at stated parts of the question. Yet the ultimate result was generally true, though the road he travelled in search of it was unknown to all but himself, and perhaps it would not have been in his power to have shown it to another." *

The statement about his taking to bed to study his more difficult problems is curiously confirmed by Brindley's own note-book, in which he occasionally enters the words "lay in bed," as if to mark the period, though he does not particularise the object of his thoughts on such occasions. It was a great misfortune for Brindley, as it must be to every man, to have his mental operations confined exclusively within the limits of his profession. Anthony Trollope well observes, that "industry is a good thing, and there is no bread so sweet as that which is eaten in the sweat of a man's brow; but the sweat that is ever running makes the bread bitter." Brindley thought and lived mechanics, and never rose above them. He found no pleasure in anything else; amusement of every kind was distasteful to him; and his first visit to the theatre, when in London, was also his last. Shut out from the humanising influence of books, and without any taste for the politer arts, his mind went on painfully grinding in the mill of mechanics. " He never seemed in his element," said his friend Bentley, "if he was not either planning or executing some great work, or conversing with his friends upon subjects of importance." To the last he was full of projects and full of work; and then the wheels of life came to a sudden stop, when he could work no longer.

It is related of him that, when dying, some eager canal undertakers insisted on having an interview with him.

---

* 'Biographia Britannica,' 2nd Ed. Edited by Dr. Kippis. The materials of the article are acknowledged to have been obtained principally from Mr. Henshall by Messrs. Wedgwood and Bentley, who wrote and published the memoir in testimony of their admiration and respect for their deceased friend, the engineer of the Grand Trunk Canal.

They had encountered a serious difficulty in the course of constructing their canal, and they *must* have the advice of Mr. Brindley on the subject. They were introduced to the apartment where he lay scarce able to gasp, yet his mind was clear. They explained their difficulty—they could not make their canal hold water. " Then puddle it," said the engineer. They explained that they had already done so. " Then puddle it again—and again." This was all he could say, and it was enough.

It remains to be added that, in his private character, Brindley commanded general respect and admiration. His integrity was inflexible ; his manner, though rough and homely, was kind ; and his conduct unimpeachable.* He was altogether unassuming and unostentatious, and dressed and lived with great plainness. His was the furthest possible from a narrow or jealous temper, and nothing gave him greater pleasure than to assist others with their inventions, and to train up a generation of engineers in his pupils, qualified to carry out the works he had himself designed, when he should be no longer able to conduct them. The principal undertakings in which he was engaged up to the time of his death were carried on by his brother-in-law, Mr. Henshall, formerly his clerk of the works on the Grand Trunk Canal, and by his able pupil, Mr. Robert Whitworth,

---

* It has, indeed, been stated in the strange publication of the last Earl of Bridgewater, to which we have already alluded, that when in the service of the Duke, Brindley was " drunken." But this is completely contradicted by the testimony of Brindley's own friends ; by the evidence of Brindley's note-book, from repeated entries in which it appears that his " ating and drink " at dinner cost no more than 8*d*. ; by the confidence generall' reposed in him, and the friendship entertained for him, by such men as Josiah Wedgwood ; and by the fact of the vast amount of work that he subsequently contrived to get through. No man of " drunken " habits could possibly have done this. We should not have referred to this topic but for the circumstance that the late Mr. Baines, of Leeds, has quoted the Earl's statement, without contradiction, in his excellent ' History of Lancashire.'

for both of whom he had a peculiar regard, and of whose integrity and abilities he had the highest opinion.

Brindley left behind him two daughters, one of whom, Susannah, married Mr. Bettington, of Bristol, merchant, afterwards the Honourable Mr. Bettington, of Brindley's Plains, Van Diemen's Land, where their descendants still live. His other daughter, Anne, died unmarried, on her passage home from Sydney, in 1838. His widow, still young, married again, and died at Longport in 1826. Brindley had the sagacity to invest a considerable portion of his savings in Grand Trunk shares, the great increase in the value of which, as well as of his colliery property at Golden Hill, enabled him to leave his family in affluent circumstances.

Before finally dismissing the subject of Brindley's canals, we may briefly allude to the influence which they exercised upon the enterprise as well as the speculation of the time. " When these fellows," says Sheridan in the ' Critic,' " have once got hold of a good thing, they do not know when to stop." This might be said of the speculative projectors of canals, as afterwards of railways. The commercial success which followed the opening of the Duke's Canal, and shortly after it the Grand Trunk, soon infected the whole country, and canal schemes were projected in great numbers for the accommodation even of the most remote and unlikely places.

In those districts where the demand for improved water communication grew out of an actual want—as, for instance, where it was necessary to open up a large coal-field for the supply of a population urgently in need of fuel—or where two large towns, such as Manchester and Liverpool, required to be provided with a more cheap and convenient means of trading intercourse than had formerly existed— or where districts carrying on extensive and various manufactures, such as Birmingham, Wolverhampton, and the Potteries, needed a more ready means of communication with other parts of the kingdom—there was no want of trade for the canals; and those constructed for such purposes very soon had as much traffic as they could carry. The owners of land discovered that their breed of horses was not destroyed, and that their estates were not so cut up as to be rendered useless, as many of them had prognosticated. On the contrary, the demand for horses to carry coals, lime, manure, and goods to and from the canal depôts, rapidly increased. The canals meandering through their green fields were no such unsightly objects after all, and they very soon found that inasmuch as the new waterways readily enabled agricultural produce to reach good markets in the large towns, they were likely even to derive considerable pecuniary advantages from their formation.

Another objection alleged against canals, on public grounds, was alike speedily disproved. It was said that inland navigation, by reason of its greater cheapness, ease, and certainty, must necessarily diminish the coasting trade, and consequently discourage the training of seamen, who formed the constitutional bulwark of the kingdom. But the extraordinarily rapid growth of the shipping-trade of Liverpool, and the vastly increased number of seagoing vessels required to accommodate the traffic converging on that seaport, very soon showed that canals, instead of diminishing, were calculated immensely to promote the naval power and resources of England. Thus it was found that in the thirty years which elapsed subsequent to the opening of the Duke's Canal between Worsley and Manchester,—during which time the navigation had also been opened to the Mersey, and the Grand Trunk and other main canals had been constructed, connecting the principal inland towns with the seaports,—the tonnage of English ships had increased threefold, and the number of sailors been more than doubled.

So great an impulse had thus been given to the industry of the country, and it had become so clear that facility of communication must be an almost unmixed good, that a desire for the extension of canals sprang up in all districts; and instead of being resisted and denounced, they became everywhere the rage. They were advocated in pamphlets, in newspapers, and at public meetings. One enthusiastic pamphleteer, advocating the formation of a canal between Kendal and Manchester, denounced the wretched state of the turnpike-roads, which were maintained by "an enormous tax," and exclaimed, " May we all scorn to plod through the dirt as we long have done at so large an expense; and for the support of our drooping manufactories, let canals be made through the whole nation as common as the public highways." *

There seemed, indeed, to be every probability that this desire would be shortly fulfilled; for so soon as the canals which had been made began to pay dividends, the strong motive of personal gain became superadded to that of public utility. The rapid increase of wealth which they promoted served to stimulate the projection of new schemes; and in a very few years after Brindley's death we find an immense number of Navigation Acts receiving the sanction of the legislature, and canal works in progress in all parts of the country. The shares were quoted upon 'Change, where they became the subject of commerce, and very shortly of wild speculation.

By the year 1792, the country was in a perfect ferment about canal shares. Notices of eighteen new canals were published in the ' Gazette' of the 18th August in that year. The current premiums on single shares in those companies for which Acts had been obtained were as follows : Grand Trunk, 350*l.*; Birmingham and Fazeley, 1170*l.*; Coventry, 350*l.*; Leicester, 155*l.*; and so on. There was a rush to secure allotments in the new schemes, and the requisite capi-

---

* ' A Cursory View of a proposed Canal from Kendal to the Duke of Bridgewater's Canal, leading to the great manufacturing town of Manchester.' 1785.

tals were at once eagerly subscribed.   At the first meeting, held in 1790, of the promoters of the Ellesmere Canal, 112 miles in extent, to connect the Mersey, the Dee, and the Severn, applications were made for four times the disposable number of shares.

A great number of worthless and merely speculative schemes were thus set on foot, which brought ruin upon many, and led to waste both of labour and capital.   But numerous sound projects were at the same time launched, and an extraordinary stimulus was given to the prosecution of measures, too long delayed, for effectually opening up the communications of the country.   The movement extended to Scotland, where the Forth and Clyde Canal, and the Crinan Canal, were projected; and to Ireland, where the Grand Canal and Royal Canal were undertaken.   But, as Arthur Young pithily remarked, in reference to these latter projects, " a history of public works in Ireland would be a history of jobs."

In the course of the four years ending in 1794, not fewer than eighty-one Canal and Navigation Acts were obtained : of these, forty-five were passed in the two latter years, authorising the expenditure of not less than 5,300,000*l.* As in the case of the railways at a subsequent period, works which might, without pressure upon the national resources, easily have been executed if spread over a longer period, were undertaken all at once ; and the usual consequences followed, of panic, depreciation, and loss.

But though individuals lost, the public were eventually the gainers.   Many projects fell through, but the greater number were commenced, and after passing through the usual financial difficulties, were finished and used for traffic.   The country became thoroughly opened up in all directions by about 2600 miles of navigable canals in England, 276 miles in Ireland, and 225 miles in Scotland. The cost of executing these great water-ways is estimated to have amounted to about fifty millions sterling.   There was not a place in England south of Durham, more than fifteen miles from water communication; and most of the large towns, especially in the manufacturing districts, were directly accommodated with the means of easy transport of their goods to the principal markets. "At the beginning of the present century," says Dr. Aiken, writing in 1795, "it was thought a most arduous task to make a high road practicable for carriages over the hills and moors which separate Yorkshire from Lancashire, and now they are pierced through by three navigable canals!"

Notwithstanding the great additional facilities for conveyance of merchandise which have been provided of late years by the construction of railways, a very large proportion of the heavy carrying trade of the country still continues to be conducted upon canals. It was indeed at one time proposed, during the railway mania, and that by a somewhat shrewd engineer, to fill up the canals and make railways of them. It was even predicted, during the construction of the Liverpool and Manchester Railway, that "within twelve months of its opening, the Bridgewater Canal would be closed, and the place of its waters be covered over with rushes." But canals have stood their ground, even against railways; and the Duke's Canal, instead of being closed, continues to carry as much traffic as ever. It has lost the conveyance of passengers by the fly-boats,* it is true; but it has retained and in many respects increased its traffic in minerals and merchandise. The canals have stood the competition of railways far more

---

* The following curious paragraph is from the 'Times' of the 19th December, 1806. It relates to the despatching of troops from London for Ireland, during a time of great excitement :— "The first division of the troops that are to proceed by Paddington Canal for Liverpool, and thence by transports for Dublin, will leave Paddington to-day, and will be followed by others to-morrow, and on Sunday. By this mode of conveyance the men will be *only seven days* in reaching Liverpool, and with comparatively little fatigue, and it would take them above fourteen days to march that distance. Relays of fresh horses for the canal-boats have been ordered to be in readiness at all the stations "

successfully than the old turnpike-roads, though these too are still, in their way, as indispensable as canals and railways themselves. Not less than twenty millions of tons of traffic are estimated to be carried annually upon the canals of England alone, and this quantity is steadily increasing. In 1835, before the opening of the London and Birmingham Railway, the through tonnage carried on the Grand Junction Canal was 310,475 tons; and in 1845, after the railway had been open for ten years, the tonnage carried on the canal had increased to 480,626 tons. At a meeting of proprietors of the Birmingham Canal Navigations, held in October, 1860, the chairman said, "the receipts for the last six months were, with one exception, the largest they had ever had."

Railways are a great invention, but in their day canals were as highly valued, and indeed quite as important; and it is fitting that the men by whom they were constructed should not be forgotten. We may be apt to think lightly of the merits and achievements of the early engineers, now that works of so much greater magnitude are accomplished without difficulty. The appliances of modern mechanics enable men of this day to dwarf by comparison the achievements of their predecessors, who had obstructions to encounter which modern engineers know nothing of. The genius of the older men now seems slow, although they were the wonder of their own age. The canal, and its barges tugged along by horses, may appear a cumbersome mode of communication, beside the railway and the locomotive with its power and speed. Yet canals still are, and will long continue to form, an essential part of our great system of commercial communicati n,—as much so as roads, railways, or the ocean itself.

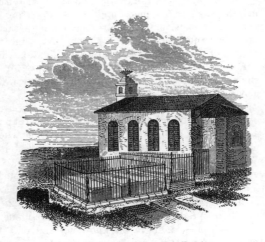

Brindley's Burial-place at New Chapel.

APPENDIX.

The Grand Canal of Languedoc and its Constructor, Pierre-Paul Riquet de Bonrepos.

The Canal du Midi, more commonly known as the Grand Canal of Languedoc, was one of the most important works of the kind at the time at which it was executed, though it has since been surpassed by many canals in France, as well as in England and other countries. It was commenced in 1666, about a hundred years before Brindley began the Bridgewater Canal, and it was finished in 1681. The magnitude and importance of the work will be understood from the following brief statement of facts.

The canal crosses the isthmus which connects France with Spain, along the valley between the Pyrenees and the Rhone, and extends from the Garonne at Toulouse (which is navigable from thence to the Bay of Biscay) to Cette on the shores of the Mediterranean, thus uniting that sea with the Atlantic Ocean. The length of the navigation, from Toulouse to Cette, is about 158 English miles, including its passage through Lake Thau, near Cette, where, in consequence of the shallowness of the lake, it is confined for a considerable distance within artificial dykes. The canal is carried along its course over rivers, and under hills, by means of numerous aqueducts, bridges, and tunnels. From the Garonne to the summit it rises 207 feet by twenty-six locks; the summit level is 3½ miles, after which it descends by thirty-seven locks into the Aude near Carcassonne. It then proceeds along the north side of that river, passing over several streams, and descending by twenty-two locks into Lake Thau. There are other locks in the neighbourhood of Toulouse and Cette; the whole number being above a hundred. The fall from the summit at Naurouse to the Mediterranean is 621½ feet; a fact of itself which bespeaks the formidable character of the undertaking.

The Grand Canal of Languedoc was constructed by Pierre-Paul Riquet de Bonrepos, a man of extraordinary force of character, bold yet prudent, enterprising and at the same time sagacious and patient, possessed by an inexhaustible capacity for work, and endowed with a faculty for business, as displayed in his organization of the labours of others, amounting to genius. Yet Riquet, like Brindley, was for the most part self-taught, and was impelled by his instincts rather than by his education to enter upon the construction of canals, which eventually became the great business of his life. Presuming that an account of the " French Brindley " will not be uninteresting to English readers, we append the following summary of Riquet's career in connection with the great enterprise in question.

The union of the Mediterranean with the Atlantic by means of a navigable canal across the South of France had long formed the subject of much curious speculation. The design of such a work will be found clearly sketched out in the ' Mémoires de Sully ;' but the project seemed to be so difficult of execution, that no steps were taken to carry it into effect until it was vigorously taken in hand by Riquet in the reign of Louis XIV. Though descended from a noble stock (the Arrighetti or Riquetti of Florence, to a branch of which Riquetti Marquis de Mirabeau belonged), Riquet, at the time he took the scheme in hand, was only a simple exciseman (homme de gabelle). His place of residence was at the village of Bonrepos, situated near the foot of the Montagne Noire, where he owned some property.

France is there at about its narrowest part, and it had naturally occurred to those who speculated on the subject of a canal, that it would be of great public importance if by such means the large navigable river, the Garonne, which flowed into the Western Ocean, could be united to the smaller river, the Aude, which flowed into the Mediterranean. Both had their sources in the Pyrenees, and in one part of their course the rivers were only about fourteen leagues apart. The idea of joining them was thus perfectly simple. The great difficulty was in the execution of the work, the levels being different, and the intervening country rocky and mountainous. The deputies from Languedoc to the States General of Paris had at various times brought the subject of the proposed canal under the notice of the Government, and engineers were even sent into the locality to report as to the frasibility of the scheme ; but the result of their examination only served to confirm the general impression which prevailed as to the impracticable character of the undertaking.

The situation of Riquet's property at the Montagne Noire probably had the effect of directing his attention to the subject of the proposed canal. Though the king had made him a tax-gatherer, nature had made him a mathematician; and his studies having taken a practical turn, he gradually went from geometry to levelling and surveying. His instruments and appliances were of the simplest sort, but they proved sufficient for his purpose. The Chancellor d'Aguesseau, in the memoir of his father, who personally knew Riquet, says of him that "the only instrument he then possessed was a wretched compass of iron; and it was this, with very little instruction and assistance, that led him, guided mainly by his powerful natural instinct, which often avails more than science, to form the daring conception of uniting the Ocean with the Mediterranean."*

He carefully examined the course of the streams in his neighbourhood, trying to find out some practicable method of uniting the Garonne and the Aude, but more especially some means of supplying the upper levels of the Canal which he had in his mind's eye with sufficient water for purposes of navigation. With those objects he made many surveys of the adjacent country, until he knew almost every foot of the ground for thirty miles round. In the mean time he hired one Pierre, the son of a well-sinker of Revel, to dig a number of experimental little canals under his direction, in his gardens at Bonrepos, where they are still to be seen. These miniature works included conduits, sluices, locks, and even a model tunnel through a hill.

At length, in the year 1662, he brought his plans under the notice of the famous Minister Colbert. Addressing him from the village of Bonrepos, Riquet said, "You will be surprised that I should address you on a matter about which I might be supposed to know nothing, and that an exciseman should mix himself up with river surveying. But you will excuse my presumption when I inform you that it is by the order of the Archbishop of Toulouse that I write to you."† He then proceeded to state that, having made a particular study of the subject, he had formed definite plans for carrying into effect the

---

* 'Œuvres de d'Aguesseau,' tome xiii.
† 'Histoire du Canal de Languedoc,' 1.

par les Descendants de Pierre-Paul Riquet de Bonrepos. Paris, 1805.

proposed canal, of which he enclosed a description, though in a very imperfect form; "for," he added, "not having learnt Greek nor Latin, and scarce knowing how to speak French, it is not possible for me to explain myself without stammering." Pointing out the great advantages to the nation of the proposed canal, the time and the money that would be saved by enabling French ships to avoid the long voyage between the west coast and the Mediterranean by the Straits of Gibraltar, while the resources of the rich districts of Languedoc and Guienne would be freely opened up to the operations of commerce, Riquet concluded by stating that when he had the pleasure of learning that the project met with the general approval of the Minister, he would send him the details of his plans, the number of locks it would be necessary to provide, together with his calculations of the exact length, breadth, depth, and other particulars of the proposed canal.

Colbert, at that time Controller-General of Finance, was actively engaged in opening up new sources of wealth to France, and Riquet's plan at once attracted his attention and excited his admiration. He lost no time in bringing it under the notice of Louis XIV., whose mind was impressed by all undertakings which bore upon them the stamp of greatness. The king saw that Riquet's enterprise, if carried out, was calculated to add to the glory of his reign; and he resolved to assist it by all the means in his power. By his order, a Royal Commission was appointed to inquire into the scheme, examine on the spot the direction of the proposed navigation, and report the result.

Meanwhile Riquet was not idle. He walked over the entire line of the intended canal several times, correcting, amending, and perfecting the details of his plans with every possible care. "I have gone everywhere, all over the ground," he wrote to the Archbishop, "with level, compass, and measuring line, so that I am perfectly acquainted with the route, the various lengths, the locks that will be required, the nature of the soil, whether rock or pasture, the elevations, and the number of mills along the line of navigation. In a word, my Lord, I am ignorant of no point of detail in the project, and the plan which I am prepared to submit will be faithful, being made on the ground, and with full knowledge of the subject." His survey finished, Riquet proceeded to Paris to see Colbert, to whom he was introduced by his friend the Archbishop; and after many conferences, Riquet returned to Languedoc to prepare for the inspection of the proposed line of navigation by the Commission. Their labours extended over two months, beginning at Toulouse and ending at Beziers.* The result of the investigation was favourable to Riquet's plan, which was pronounced practicable, with certain modifications, more particularly as respected the water-supply; and the Commissioners further recommended the extension of the proposed canal to a harbour at Cette.

A long correspondence ensued between Riquet and Colbert as to the details of the scheme. Riquet disputed the conclusions of the Commissioners as to the alleged difficulty of constructing the works near Pierre de Naurouse intended for the supply of water to the canal; and, to show his confidence in his own plans, he boldly offered to construct the conduit there at his own risk and cost: "in undertaking to do which," said he, "I risk both honour and goods; for if I fail in satisfactorily performing the work, I shall pass for a mere visionary, and at the same time lose a considerable sum of money." So much difference of opinion prevailed upon this point,—the men of science alleging the inadequacy of Riquet's plans, and Riquet himself urging their perfect sufficiency,—that Colbert resolved that until this point was decided, no further step should be taken to authorize the canal itself to be proceeded with.

But in order that Riquet might have an opportunity of vindicating the soundness of his plans, and not improbably with a view to test his practical ability to carry out the larger works of excavation and construction, letters patent were issued entrusting him to proceed with the necessary trench or conduit to enable some experience to be obtained of the inclination of the ground and the probability or otherwise of obtaining an adequate supply of water for the navigation.

Riquet, with his usual promptitude, immediately began the execution of this work. The rapidity with which he proceeded surprised everybody; and the conduit was very speedily finished to the entire satisfaction of the Government inspectors. Having given this proof of his practical skill, and demonstrated the possibility of supplying sufficient water to the summit level, the king determined on authorizing Riquet to proceed with the construction of the canal.

The question of providing the requisite money for the purpose formed the next important subject of consideration. As the province of Languedoc would derive the principal advantages from the navigation being opened out, it was proposed at the assembly of the

---

* The Procès-Verbal des Commissaires is preserved in the Manuscrits Colbert, No. 202, Bibliothèque Imperiale.

States, in 1665, that that province should contribute a certain proportion of the cost, on condition that the king should provide the remainder. Notwithstanding, however, the great advantages to the province of the construction of the canal, the States of Languedoc would not untie their purse-strings; and they declared (26th February, 1666) that they would neither then nor thereafter contribute towards its cost. On the other hand, the royal treasury had become almost exhausted, and any new expense could with difficulty be borne by it.

Riquet's demonstration of the possibility of uniting the Mediterranean with the ocean had therefore thus far been in vain. But he had fairly committed himself to the enterprise, and he was not the man now to turn back. Having had the courage to conceive the design, he urged the Government to allow the work to proceed; and he suggested a method by which in his opinion the means might be provided without unnecessarily burdening the public treasury. He offered to construct the works upon the first division of the canal, from Toulouse to Trebes near the river Aude, within a period of eight years, for the sum of 3,630,000 livres; and in payment of this sum he proposed to the king that he should grant him the sole farming of the taxes in Languedoc, Roussillon, Conflans, and Cerdagne during six years, on the same terms at which they were then held, and also that the offices of Controllers of peasants'-tax (des tailles), of tradesmen's-tax, and of rights over the salt-works of Pecais, should at the same time be assigned to him. This was a bold offer; but the Council of State accepted it, and the necessary legal powers were accordingly granted to Riquet, upon which measures were taken to enable him to proceed with the works.

The plans which Riquet had prepared were revised and settled by M. Cavalier, the Commissaire-Général of Fortifications, who acted on behalf of the king. Riquet's plans had been prepared with such care, that in his specification Cavalier for the most part adopted them, only altering the dimensions,* which were somewhat increased. The King's Engineer was careful not to define too rigidly the manner in which the works were to be carried out. He did not even trace a definite line to be followed, but merely marked the general limits within which the canal was to run. He foresaw that, with experience, many modifications might be found necessary

---

* 'Canaux Navigables,' par La Lande, 120.

in the progress of the works, and he preferred leaving the management of the details as much as possible to the skill and judgment of the contracting engineer.*

Riquet had now to display his genius in a new sphere. Hitherto he had exhibited himself mainly as a designer; he had drawn plans, advocated and explained them to others, and transacted the part of a diplomatist in getting them adopted. Though his execution of the conduit near Pierre de Naurouse had enabled him to give satisfactory proofs of his engineering skill, the work he was now about to enter upon was of a much more formidable character, calling for the exercise of a varied class of practical qualities. He had to direct the labours of a very large number of men, to select the most suitable persons to superintend their various operations, and meanwhile to give his continuous attention to the carrying out of his plans, which, as in the case of all great undertakings, required constant modification according to the many unforeseen circumstances which from time to time occurred in the course of their execution.

Anxious to proceed with the greatest despatch, Riquet began with the organization of his staff of workmen and superintendents. He divided them into a number of distinct groups, appointing a chef d'attelier to each, under whom were five brigadiers, each brigadier having the direction of fifty workmen. These groups were again combined in departments, a controller-general being appointed over each, and under him were travelling controllers, who received the reports of the brigadiers and chefs d'atteliers, and thus effectually maintained and concentrated the operations of the workpeople, who sometimes numbered from eleven to twelve thousand.

Before Riquet had proceeded far with his enterprise, he experienced a difficulty which has proved the bane of many a grand scheme—want of money! The produce of the taxes above referred to was not sufficient to enable him to push on the works with vigour; but, rather than they should be delayed, he himself incurred heavy debts, selling or mortgaging all his available property to raise the requisite means. Notwithstanding the early decision of the States of Languedoc not to contribute towards the cost of constructing the canal, Riquet again and again made urgent appeals to them for help, but for some time in vain. Several grants had been made from the royal treasury to enable the contract to proceed, but Louis XIV. having become engaged in one of his expensive wars, was no longer able to

---

* 'Histoire du Canal de Languedoc,' p. 50.

contribute; and Riquet, having come to the end of his own resources, began to fear lest the canal works should be brought to a complete standstill.

Colbert continued his fast friend and supporter, and took the most lively interest in the prosecution of the undertaking. His name was indeed a tower of strength, and his influence was in itself equal to a large capital. Of this Riquet on one occasion made very adroit use, for the purpose of inducing the States of Languedoc at length to take part in his enterprise. It is said that, in order to impress upon their minds the confidence reposed in him by the great Minister, and thus to enhance his credit with them, he persuaded Colbert to allow him to try the following ruse. He asked to be permitted to enter his cabinet while he was engaged with the farmers-general of the province in discussing the renewal of their leases. Colbert consented; and while thus occupied, Riquet turned the key of the room-door, and entered and sat down, without saying a word to any one, and without any one speaking to him. The farmers-general looked at Riquet, then at the Minister, who took no heed of him, and then at each other. Strange that Colbert should place such confidence in Riquet as to permit him thus to enter his private chamber at pleasure! A second meeting of the same kind took place, and again Riquet entered as before. After the interview was over, the farmers spoke to Riquet of his canal, and its great utility; and ended by offering to lend him 200,000 livres. Riquet, however, listened to the proposal very coolly, without accepting it. At the end of a third interview with the Minister, at which Riquet was present as before, the farmers raised their terms, and offered to lend him 500,000 livres. Riquet replied that he could do nothing without the sanction of the Minister; and re-entering the cabinet of Colbert, he related to him what had passed. The Minister was very much amused at Riquet's adroitness, and readily gave his sanction to the proposed loan.*

This advance proved the beginning of a series of loans of great magnitude advanced to Riquet by the States of Languedoc to enable him to complete the canal. Though they were slow to believe in the practicability of the scheme, and for some time regarded it as impossible, their views changed when they saw the first part of the canal completed from Toulouse to Trebes, and opened for the purposes of navigation. They were then ready to recognize its great public uses, and from that time forward they exerted themselves to raise the neces-

---

* This anecdote is given in the ' Mémoires du Baron de Bésenval.'

sary money to enable it to be completed to its junction with the Mediterranean at the port of Cette.

But Riquet had numerous difficulties to encounter besides those arising from want of money in the course of his undertaking. The prosecution of the works involved constant anxiety and unremitting labour. One of his greatest troubles was the conciliation of the owners of the land through which the canal passed, many of whom were extremely hostile to the enterprise, and feared that it would inflict irreparable injury upon their property. Worse than all was the malice, misrepresentation, and calumny that pursued him. He was denounced as an impostor, attempting to do an impossible thing; he was wasting public money upon a work which, even if finished, could never be of any use. Before he began the canal, it was predicted that water enough could never be collected to supply the summit level; and after that difficulty had been satisfactorily solved, local detraction was directed against the works. " Indeed," writes M. de Froidour to M. de Barillon, " the people of the locality are so agreed in decrying them that the wonder is to find a person who has not arrived at the foregone conclusion that this enterprise can never succeed."*

Notwithstanding the alleged uselessness of the canal, the insufficiency of the works, and the prophecies of its failure even if completed, Riquet bravely bore up, amidst toil, and disappointments, and bodily suffering. He never lost hope or courage, but persevered through all. Writing to Colbert in April, 1667, he said, " I now know the strong as well as the weak points of my work better than I before knew them; and I can assure you in all truth that it would not be easy to imagine a grander or more useful undertaking; I have all the water-supply that I require, and the invention of my reservoirs will furnish me during summer with sufficient to render the navigation perpetual."

At another time he wrote to Colbert :—" My enterprise is the dearest of my offspring : I look chiefly to the glory of it, and your satisfaction, not my profit ; for though I wish to leave the honour to my children, I have no ambition to leave them great wealth." And again—" My object is not to enrich myself, but to accomplish a useful work, and prove the soundness of my design, which most people have hitherto regarded as impossible."

---

* Three Letters from M. de Froidour to M. de Barillon, published at Tou louse in 1672.

About the beginning of 1670, after three years' labour, part of the canal was opened, from Toulouse to Duperier, and used for the transport of materials. This was a comparatively easy section of the undertaking. But Riquet was desirous of exhibiting the practical uses of the canal at the earliest possible period, not only for the purpose of mitigating the popular opposition, but of encouraging the king, Colbert, and the States-General, to support him with the necessary means to complete the remainder of the navigation from Trebes to Cette. Two years later, a further portion was finished and opened for public traffic. The Archbishop embarked at Naurouse, and sailed along the new canal down to Toulouse. Four large barques ascended from the Garonne to Naurouse, and returned freighted with provisions and merchandise. The merchants of Gaillac, who had before been unable to send their wines to Bordeaux for sale, were now enabled to do so, and they established a packet-boat which regularly went between Naurouse and Toulouse three times a week.

The remaining portions of the canal were in full progress. The basins, the conduits, the locks, were all well advanced as far as Castelnaudary, and Riquet was vigorously grappling with and successively overcoming the great difficulties which occurred in the construction of the works between that place and the Mediterranean. Not the least among the number of his obstructions were the quarrels between the two Commissioners appointed by the king on the one hand, and by the States of Languedoc on the other, to superintend the execution of the undertaking. Each represented particular local interests, and whilst one desired to keep the canal to the north, the other wished it to proceed more to the south by way of Narbonne. Between their contentions, Riquet had sometimes a difficult course to steer. Thus, at Malpas, where it was necessary to carry the canal by a tunnel under the hill of Enserune, both Commissioners pronounced the work to be impracticable, because the hill appeared to consist chiefly of a sandy stuff permeable to water, and apt to give way. But they were far from agreed as to the remedy, each urging upon Riquet the adoption of an opposite course, one that he should carry the canal northward by Maureillan, the other that he should carry it southward by Nissau and Vendres. They both wrote to Colbert, pointing out that Riquet's plan could never be carried out; that the scheme threatened to prove a total failure, because his work had run its head into a sand-hill, at a point where it had a lake on each side of it, from twenty-five to thirty feet below its level. In the mean time the Commissioners gave Riquet orders to suspend the further prosecution of the works.

Riquet put the orders in his pocket, and coolly resolved to carry out his own plan. To conceal his design, he pretended to abandon the trench leading to the mountain, and sent the workmen to that part of the canal which lay between Beziers and Agde. He then privily set a number of excavators to work upon the mountain-side near Malpas, and in six days he had vanquished the "impossibility," and cut a clear passage through it for his canal! When the work was finished, he sent invitations to the Cardinal de Bonzy and to the two Commissioners to come and inspect what he had done. Greatly to their surprise, he led them right through his tunnel, lit up with flambeaux; and his triumph was complete.

It was not so easy to overcome the constantly recurring difficulties occasioned by the want of money. Nothing but money would satisfy his thousands of workmen and workwomen (for of the latter about six hundred were employed), and he was often put to the greatest straits for want of it. We find him, in 1675, writing to Colbert in very urgent terms. Unless supplied with funds, he represented that it would be impossible for him to go on. "People tell me," said he, "that I am only digging a canal in which to drown myself and my family." The king ordered a remittance to be sent to Riquet, but it was insufficient for his purposes. The costly harbour works at Cette were now in full progress, absorbing a great deal of money; and Riquet's creditors grew more and more clamorous. Colbert urged the States of Languedoc to give him more substantial help, and they complied so far as to vote him a loan of 300,000 livres. But this money was only to be advanced at different and remote periods. Thus it did little to help him either in credit or in funds. "I am doing everything that is possible," he again wrote to Colbert, on the 21st of January, 1679, "to find persons that will lend me money to enable me to finish the canal in the course of the present year. But I am so overwhelmed with debt, that nobody will trust me, so that I am under the necessity of again having recourse to you, and informing you of my needs; you will see what they are by the enclosed statement; I venture to ask that you will state your wishes alongside of each item, so that I may be put in a position to bring my enterprise to a successful termination: it is my passion, and I shall be in despair if I cannot finish it. Time flies; and once lost, it can never be recovered."

Riquet had indeed reason to be apprehensive that he might not live to complete his work. The strain upon his mind and body for the fifteen years during which the canal was in progress had been very great, and he had more than one serious attack of illness. But still the enterprise went forward. The organization established by him was so perfect that his occasional absence from this cause was scarcely felt; besides, his eldest son was now in a manner competent to supply his place.

The third portion of the undertaking, consisting of the harbour and sea entrance to the canal at Cette, was being vigorously pushed on, and the canal was almost ready for opening throughout from one end of it to the other, when, worn out by toil and disease, Riquet breathed his last, without having the satisfaction of seeing his glorious work brought to completion. The canal was finished under the superintendence of his son, and was opened for public traffic about six months after Riquet's death.

The total cost of the canal was about sixteen million livres, equal to 1,320,000*l.* sterling. Riquet sank all his own means in it, and when he died it was found that his debts amounted to more than two million livres. To defray them, his representatives sold the principal part of the property he held in the concern, and it was not until the year 1724, forty years after the navigation had been opened, that it began to yield a revenue to Riquet's heirs.

As in the case of many other great works, Riquet's merits have not been left undisputed. Thirty years after his death, a claim was set up on behalf of one of his assistants, M. Andreossy, as having been the designer of the canal; while, more recently, a like claim has been made on behalf of M. Cavalier, the king's engineer. Although all contemporary witnesses had died before Riquet's merits, till then undisputed, were thus called in question, happily the Archives of the canal as well as the Colbert correspondence survive to prove that, beyond the shadow of a doubt, Riquet was not only the inventor and the designer, but the constructor, of the Grand Canal of Languedoc.

It must, however, be admitted that the genius of this great engineer has received but slight recognition among his own countrymen, there not being a single tolerable memoir of Riquet to be found in the whole range of French literature. To this, however, it might be rejoined, that only a few years since the same thing might have been said of our own Brindley.

INDEX.

THE END.